A
DICTIONARY
OF
MACHINING

By the same author:

CEMENT AND CONCRETE ENGINEERING
CUTTING TOOL MATERIALS
DICTIONARY OF ALLOYS
DICTIONARY OF FERROUS METALS
DICTIONARY OF FOUNDRY WORK
DICTIONARY OF HEAT TREATMENT
THE GRINDING OF STEEL
A GUIDE TO UNCOMMON METALS
IRON—SIMPLY EXPLAINED
MACHINE TOOLS
MARKETING THE TECHNICAL PRODUCT
MECHANICS FOR THE HOME STUDENT
METAL FATIGUE
METAL WEAR
AN OUTLINE OF METALLURGY
SAWS AND SAWING MACHINERY
STEEL CASTINGS
STEEL FILES AND THEIR USES
STORY OF METALS
SUCCESSFUL BUYING
SUCCESSFUL RETAILING
SURFACE TREATMENT OF STEEL
TESTING OF METALS
WELDING

In Collaboration:

FOUNDRY PRACTICE
HEAT-TREATMENT OF STEEL
MACHINING OF STEEL
MECHANICAL WORKING OF STEEL
STAINLESS AND HEAT-RESISTING STEELS
STEEL MANUFACTURE—SIMPLY EXPLAINED
STEEL WORKING PROCESSES
STORY OF THE SAW
STRUCTURE OF STEEL
WELDING PROCESSES AND TECHNOLOGY

A DICTIONARY OF MACHINING

Eric N. Simons

FREDERICK MULLER LIMITED

First published in Great Britain in 1972 by
Frederick Muller Limited, 110 Fleet Street, London EC4A 2AP

Copyright © 1972 Eric N. Simons
Appendix copyright © 1972 Paul Barnett

All rights reserved. No part of this publication may be reproduced, stored in a retrieval system, or transmitted, in any form or by any means, electronic, mechanical, photocopying, recording or otherwise, without the prior permission of Frederick Muller Limited.

Ac. No. 1975.9

621.75

Printed in Great Britain by The Anchor Press Ltd.,
and bound by William Brendon & Son Ltd.,
both of Tiptree, Essex

SBN 584 10067 1

To
My Son, Martin Simons,
M.Ed., B.Sc.

Contents

	page
Index of Illustrations	7
Preface	9
A Dictionary of Machining	13
Appendix: Theory of Machining Outlined	233

Index of Illustrations

	figures	page
Acme screw thread	1	14
Adjustable reamer	42	155
Bandsaw (vertical)	2	17
Bandsaw tooth arrangement	44	166
Bandsawing machine (horizontal)	3	17
Broaching machine, surface	4	24
Buttress thread	5	27
Butt-welded tool blank	37	153
Centreless grinding	7	31
Chuck, use of the	8	41
Chucking reamer	41	155
Circular sawing machine	9	43
Concave cutter	27	129
Contour sawing machine	10	48
Cutting fluid, correct and incorrect use of	11, 12	56
Cutting-tool rake angles	58	216
Double-angle gear cutter	17	91
Drilling machine, radial	36	151
Electrical discharge machining	6	29
End mill	28	130
Equal-angle gear cutter	18	91
Face cutter	25	128
Formed cutters	30, 31	130
Gap bed lathe	14	90
Hacksaw blades	45–8	167
Hand reamer	40	155
Horizontal bandsawing machine	3	17
Involute gear cutter	15	91
Lathe and its component parts	19	119
Lathe, gap bed	14	90
Lathe, normal traverse motion of	20	120
Lathe tailstock	55	203

ILLUSTRATIONS

	figures	page
Micrometer gauge	21	124
Milling machine of vertical arbor type	22	126
Plain milling cutter	23	127
Planing machine	32	138
Plug chamfer tap	33	141
Pull broach	34	147
Push broach	35	147
Radial drilling machine	36	151
Raker and G-set contour saw teeth	50	173
Rose-shell reamer	39	155
Saw teeth	43	165
Saw teeth, segmental	49	171
Sawing machine, circular	9	43
Sawing machine, contour	10	48
Segmental saw teeth	49	171
Shaping machine	51	174
Shell end mill	29	130
Shell reamer	38	154
Side cutter	24	127
Side relief angle	52	180
Single-angle gear cutter	16	91
Single-point cutting-tool shapes	53	182–3
Slitting saw	54	187
Slotting cutter	26	128
Surface broaching machine (vertical)	4	24
Tailstock	55	203
Tool cutting angles	56, 57	215
Turning operations	59	222
Twist drill nomenclature	13	68
Vertical bandsaw	2	17

Preface

THE pace of technical progress is tremendous, even terrifying. This makes it essential, ever and again, to examine a particular technical field and present the results of this examination in a condensed but helpful form to those not always familiar with the latest developments.

This dictionary seeks to bring up to date the terminology of machining by revising previous definitions and including processes, machines and even tools not available or known when similar, less exhaustive, compilations were devised. Inevitably, for reasons of economy and space, this is a selective dictionary. It could not be otherwise. Nevertheless, such omissions as there are—and there are not many—relate less to the more recent and rather to the old-established, well-known items which it would be absurd to re-define, since a new definition would add little or nothing.

Instead, attention has been paid to giving the reader facts and data explanatory of the subjects covered so that, instead of reading "Twist Drill: a tool for boring holes in metals", he also learns something of drill design, cutting-angles, functions, etc. Without being an encyclopaedia, this might be termed an "encyclopaedic dictionary" of machining, though not of general engineering.

As examples of new entries one may mention Abrasive Jet Machining, Ceramic Tools, Chemical Machining, Chemical Contour Machining, Damper Plugs, Electrical Discharge Grinding, Electrochemical Machining, Electron Beam Machining, Formate Cutting, Helixform Cutting, Laser Beam Machining, Plasma Arc Machining, Revacycle Cutting, Spline Rolling, Tantalum Carbide, Ultrasonic Machining, etc., etc.

Illustrations are included to outline the principles of the tools and the machines concerned, and thus in many cases do not include modern refinements and details that serve, by way of explanation, merely to cloud the issue. Details of tools and machinery change swiftly, their principles more slowly, if at all. Three of the illustrations are reproduced from *Engineering and Its Language* by Bedrich Scharf, B.A., F.I.L., and I would like to take this opportunity of thanking him for his kind permission. I would also like to thank

PREFACE

Dr. A. A. Smith of King's College, London, for his constructive and illuminating criticisms.

It is hoped and believed that this dictionary will be of value to machine-operators, students, foremen, engineers, lecturers and instructors, metallurgists, works managers, and all concerned directly or indirectly with the machining and shaping of metals.

Eric N. Simons

A Dictionary of Machining

A

Abrader See **Rotary Abrader**.

Abrasive Belt Grinding Grinding and polishing of flat and contoured surfaces with a continuous belt previously coated with an abrasive substance, usually composed of aluminium oxide grit in various sizes, silicon carbide, or other abrasive. The belt, carried by a wheel with which it makes contact, is of steel, rubber, compressed canvas, rubber-coated canvas, solid section canvas, polyurethane or inflated rubber. The wheel runs at between 3,000–3,500 s.f.p.m. for heavy stock removal and 4,500–7,500 for polishing. Higher speeds are not recommended. The belt is lubricated to improve or regulate surface finish and prevent overheating. Mineral oil and emulsions of soluble oil and water are usually recommended for this.

Abrasive Cutting The cutting off of sections from a metal bar by an abrasive wheel composed of aluminium oxide, resinoid-bonded or rubber-bonded, these bonds being used respectively for dry and wet cutting when the operation is more economical than sawing.

Abrasive Honing Sticks Several hundreds of kinds of honing sticks in numerous grits and grades have been produced, rendering the truing of the usual honing stones no longer necessary. The sticks are located lengthwise on the circumference of a metallic part such as a cylinder bore, while means of expanding the sticks, adjusted to suit the cylinder walls, hold them under pressure against the wall. The tool itself is revolved and reciprocated simultaneously, and is used in a hone of positive expanding type. In Britain the sticks are more usually termed **stones**. (See **Honing**.)

Abrasive Jet Machining The cutting of thin material or material highly sensitive to heat of an injurious kind, or the producing of complicated holes more easily than with alternative methods, by discharge of abrasive particles driven against a workpiece by the flow of gas from a suitable nozzle, to eliminate surplus metal, etc. The abrasion-resistant nozzles are usually of tungsten carbide or synthetic sapphire, the former being either round, rectangular or square, the latter round only. The abrasive powders are of aluminium oxide, silicon carbide, calcium magnesium carbonate, sodium bicarbonate or small beads of glass. Aluminium oxide is for the general run of work, silicon carbide for quicker cutting of hard substances, sodium bicarbonate, suitably treated, for specially fine cleaning and etching, calcium magnesium carbonate for etching and light cleaning, glass beads for light polishing and fine deburring. The abrasive is carried at high velocity by a hose from the mixing container to the nozzle held in a fixture, with the work moving or moved by appropriate mechanisms over a stationary workpiece.

Abrasive Slurry See **Ultrasonic Machining.**

Acme Screw Threads Square threads modified to give a profile angle of 29 deg. (Fig. 1). They have greater strength than the normal square threads and are more readily machined out because

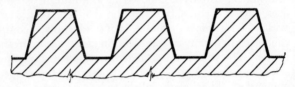

Fig. 1   Acme screw thread

of their sloping sides. Their use eliminates backlash caused by wear of the inclined sides, and their form allows a disengaging or split nut to be used. Consequently they are mostly employed to transmit motion and power. The thread forms are normally produced in the threading tool and reproduced on the work, the first 3 or 4 teeth being modified to enable the cutting power to operate over a greater area, so giving improved surface finish and longer tool life on hard materials. The first thread is cut oversize and succeeding threads are machined to the proper size, especially in the machining of such metals as high nickel copper alloy.

Adjustable Taps See **Taps.**

Air Gauge A gauge used in **honing** or **lapping** (q.v.). It is introduced into the bore with enough clearance to allow the air under pressure to pass through openings in the gauge into the outer air. With increasing bore size the back pressure in the air system declines. A carefully calibrated switch controls the pressure and stops the operation as soon as the bore attains the required size. If necessary, differing diameters can be honed without cutting off the cycle, and the gauge is specially suitable for honing the bores of cylinders in multispindle honing machines.

In lapping valve needles to close limits an air gauge is often used to enable the parts to be measured, the cycle being arrested for this purpose.

Arbour A revolving cylindrical metallic piece embodied in a machine tool. It may be the driving shaft that carries the workpiece in a lathe to enable machining to be done on the work surface, but is more usually a spindle on which are mounted cutting tools such as boring tools, milling cutters, reamers, etc. In these instances the end of the arbour is inserted into a taper socket in the machine spindle. Arbours are used in machining gears, multiple operation machining, reaming and shell tapping, etc.

Arbour Supports In some operations, such as milling, an additional support is given to the arbour to eliminate or minimize deflection, which is likely to set up excessive chipping of the teeth of tungsten carbide cutters when overloading takes place. Deflection is not wholly eliminated by this method, but staggering the cutters, adopted in conjunction with the added supports, eliminates deflection completely.

Automatics Machines introduced to carry out a wide range of machining operations automatically so that the machinist has only to introduce the work to the machine and remove it when finished. There is a considerable range of these machines of different design and construction according to the operations to be performed. They mostly embody a camshaft to regulate the succession of machining operations, this shaft being timed so that a single revolution enables every operation to be carried out and the work completed, but some automatics have a tailstock to carry one length of bar material, multiple tools being used in front and rear tool posts, working simultaneously and quickly reversing when the machine comes to a halt.

Operations carried out by automatics include broaching, chucking, burnishing, boring, sawing, die-threading, drilling, gear-cutting, milling, multiple operation machining, tapping, thread rolling, turning, flue milling, hobbing, indexing, facing, etc.

Axial Control A form of automatic control used in end-milling to extend the life of the cutting tool, give better size control, surface finish, removal of metal and reproducibility. It is particularly beneficial in machining intricate parts in aircraft, space parts and small intricate parts of instruments, etc. In end-milling the control enables the cutter to be located along the x, y and z axes.

Axial Rake Angles Angles of milling cutters measured between the circumferential cutting edge and the cutter axis when studied radially from the intersection point. A helical tooth cutter has an axial rake angle equal to the helix angle.

B

Backcoating In **chemical machining** (q.v.) the provision of connecting lines on a single mirror image master of a pair so that etching can be done on one side of the work at these positions. Thus, it holds the finished parts similarly together.

Back Rake An American term for **Negative Rake** (q.v.).

Back Spotfacers In **spot facing** (q.v.) the form of the work may render impracticable the down stroke of the machine spindle. In such instances the back spotfacer is used by locking it on the driving shaft, the spindle being revolved and feeding *upwards* into the work, the tool driver having been previously gripped in the spindle and passed through the already drilled hole.

Balance Tools Tools principally employed for taking rough cuts in turning, and held in a tool holder at an unvarying angle carefully chosen in advance of grinding the cutting edge. They do not give any tool support and are unsuitable for work of long, slim character. When a stronger cutting edge is needed, the front relief is lessened. They are much used in machining nickel alloys.

Ball Broaching Tool A special tool for burnishing the bores of bearings when installed in their housing, especially in parts made of powdered metals where the clearance between shaft and bearing is not more than ± 0.0005 in.

Ball Turning Rests Attachments in the turning of metals that take the place of the compound rest or slide (q.v.) and render feasible the turning or boring of spherical forms.

Bandsawing Cutting metals by means of a ribbon or strip of hardened steel with a toothed upper edge and carried by a pair of wheels whose diameter shows the dimension of the saw. The work is located on a work table and forced against the saw cutting edge, the work thrust being supported by roller guides held as near as possible to the work. The bandsaw can be used for cutting either wood or metal, being toothed and dimensioned accordingly. The steel saw is hardened, tempered and ground and its ends are united by butt-welding or brazing so that it forms an unbroken loop. The wheels or pulleys are often faced with rubber and the band travels in one direction only, sawing as it rotates, the work being fed towards it. The saw teeth are sharpened and set, and the steel hammered by men of great skill and experience to give tension and flatness.

BANDSAWING

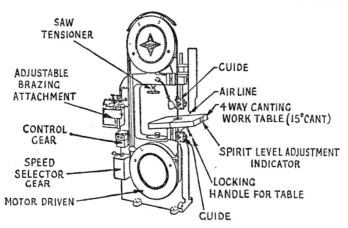

Fig. 2 Vertical bandsaw

The sawing machines may be either vertical or horizontal. Metal-cutting bandsaws are usually $\frac{1}{2}$ to 3 in. wide, and are capable of cutting aluminium alloys, magnesium alloys, copper alloys, stainless steels, titanium alloys, tool steels, beryllium, cast iron, heat-

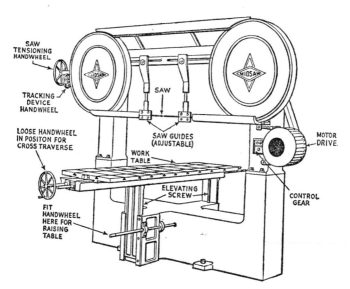

Fig. 3 Horizontal bandsawing machine

resisting alloys, malleable and ductile irons, nickel alloys, refractory metals, titanium alloys, etc.

Bar Gauge A gauge made up of a pair of bars secured to a split ring and pressed by a spring against the surface of a bore. The diameter of the bore is represented by the distance separating the bar surfaces of contact. As the tool is introduced into the bore, the bars embodied in its construction are forced inwards so that a pair of low potential electrical contacts on the split ring are kept open. With enlargement of the bore these contacts approach each other more closely, and when they meet, the cycle is halted. The gauge is primarily applied to bores more than 2 in. dia., but can be used for honed diameters within a control dimension of 0·0003 in., while adjustments within 0·040 in. can be accepted.

Bar Machines See **Automatics.** These machines are for boring, milling, multiple operation machining and reaming.

Bell Mouth A type of defect encountered in drilling metals when the twist drill produces a hole lacking perfect roundness or circularity, probably owing to defective grinding of the drill flutes. The same defect is also encountered in the honing of bores owing to the inadequate length of tools and stones.

Bell-type Drill and Countersink A high speed steel tool used in a 14-station horizontal drilling machine for machining compressor rotors. Its machine location may be on the first or second centre-drill shaft-end of the rotor. The station may be a single-spindle, cam-driven unit mounted in the centre of the work table.

Belt Grinding See **Abrasive Belt Grinding.**

Belt Polishing See **Abrasive Belt Grinding.**

Bent Shank Tapper Taps A type of taps (q.v.) employed for tapping nuts in an automatic machine. The nuts are usually contained in a hopper and fed to the machine, tapped, made to travel over the shank of the tap, and then automatically discharged over the bend end. This makes it needless to reverse the tap, as would be essential with the ordinary type. The tap shank is bent to an angle of 90 deg. for about a third of its length at the end.

Bevel Gauge A tool for accurately laying out angles. A vernier (q.v.) caliper is set to twice the sine of half the required angle, multiplied by 10, with $\frac{1}{2}$ in. added, and then the gauge is opened to fit the vernier. In this way the angle is given within the measuring tool limits and those of the gauge radius. To lay out an angle from a specific centre, a $\frac{1}{8}$ in. hole in the centre receives a setting plug. Tables for setting tools for angles and gauge settings for holes in a circle are used in conjunction with this tool.

Bevel Gear Planing Machine A machine designed for cutting teeth in a straight bevel gear by template machining, a low production, non-generating method of cutting the tooth profiles. See **Template.**

Bonded Abrasive Laps A type of lapping tool in which two laps revolve and are cooled by paraffin or other lubricating medium designed to remove chips or surplus abrasive material. The laps are operated at higher speed than those used for cast iron, have a heavier lapping action, and in consequence are incapable of maximum accuracy and do not possess the flatness of cast iron laps owing to their inability to be dressed by diamond tools. They are rigidly held on spindles and independently driven, so that the number of components to be lapped is less critical than with machines for lapping cast iron. In fact, the laps can be used satisfactorily for as few as 3 parts.

The bonded abrasive lapping machines will lap each side of a part simultaneously. Although diamond tools cannot be used for dressing them, these laps are kept flat by a special diamond-dressing attachment.

Boring Machines Machines used for boring, enlarging or completing holes in materials, but in addition for a range of work not consisting merely of producing holes in simple form with plain cutting tools. They will also produce blind holes, i.e. holes penetrating into a material, and then stopping before the tool breaks through; bottle-shaped holes; circular formed recesses and holes with a range of "steps"; counterbores and holes in a cylinder whose ends are bored to a larger size, but are parallel in form. Another operation for which they are suitable is **undercutting** (q.v.) or cutting at the side of a recess to give the machined surface an angle less than 90 deg. with the previous surface of the component. Holes may be bored much longer than the diameter of the tool, and assuming proper support is given to the work, holes whose diameter is greater than their length by as much as a factor of 50, and conversely, can be drilled.

Boring machines can be employed also in combination with turning, facing, drilling, reaming, milling, tapping, grinding and honing, but here we are concerned solely with their boring function. The machines are mainly confined to the completion of holes previously produced in cast or forged parts by coring, drilling or piercing. In modern workshops they are electronically (numerically) controlled. (See **Numerical Controls.**)

The principal machines used include engine lathes, bar machines, vertical boring mills, vertical turret lathes, horizontal boring mills, drill presses, precision-boring machines and machines of special character. These should be studied under their individual headings in this volume. The tools used in these machines are described under the heading **Boring Tools**.

Boring Mills See **Boring Machines**.

Boring Tools The least complicated boring tool is a single point cutting tool fastened firmly to a straight **boring bar**, which is rotated and fed into the work. Alternatively, the work may be made to revolve and feed against a stationary boring bar. The disadvantage of this type of tool is that as soon as the cutting edge wears or becomes dull the tool has to be taken out and reground, and must be repositioned when put back into the machine, both of which operations demand time, skill and care. For this reason some tools of this type are provided with an adjusting screw which allows the tool to move forward to allow for the wear; the holding screws are slackened and the adjusting screw turned in order to advance the tool.

Another boring tool is the **box tool** or **universal head,** which is fastened firmly to the end of the boring bar to grip various forms of cutting tools and can be employed for work involving multiple diameters. The **stub boring bar** carries a head holding a fixed cutting tool, but though uncomplicated, the tool can be adopted for only a narrow range of bore diameters. It is, however, popular and extensively used. Where a wider range of operations is required, a detachable head capable of holding a plurality of cutting tools is secured to one end of the boring bar or stub boring bar or to any point along their length. These heads are exceptionally flexible, and some are interchangeable so that the bar to which they are secured can be employed for a number of bore diameters.

The **blade tools** are of various types and designs. One has a pair of identical cutting inserts separated from each other by an angle of 180 deg. In another, the cutting tool is passed through the body to give a pair of cutting edges. Various forms of head are also adopted, such as those of multiple diameter with two or more inserts with cutting edges, so that the head may carry out a number of operations at one and the same time or in succession. In both these types of head there is no compensation for wear, but the insets, if mechanically held, may be indexed (q.v.) to keep to the size. The insets are usually of tungsten carbide, and by indexing they are enabled to utilize each cutting edge before it becomes necessary to remove and renew the tools. These multiple diameter heads are primarily intended for boring operations requiring maximum output.

Offset boring heads enable small holes to be bored, but have usually no provision for fine adjustment, unlike another type where this is sensitively given by a micro-adjustment device for use where small quantities of parts have to be bored and several modifications of diameter are involved. Radii can be generated by a special type of head used in a lathe, manually fed and employed for low production boring, but power feed can be applied for higher production. Lastly there is a head for boring at an angle of 90 deg. to the boring bar axis. This reduces vibration such as is experienced when long boring bars are used, and is particularly advantageous for producing half-bores.

Bottoming Chamfer In **tapping,** a type of **chamfer** given to solid taps required to work in blind holes in difficult materials. It is used to finish tapping previously carried out with taper chamfer or plug chamfer taps, and its advantage is that it minimizes the period during which the threads are called upon to withstand the greatest stress.

Bottoming Tap The third or "plug" tap in a standard series of three.

Bottoming Tools Tools designed to clear the bottom of a hole or to provide a suitable bottom. They must have a correct side clearance angle, and the end relief angle must be adequate to clear the bore surface, increasing as bore diameter decreases, but must not be so great that the tool edge will become insufficiently strong. (See **Tool Angles.**)

Box Chuck A type of **chuck** (q.v.) designed to enable a component, of a form such that all the machinable surfaces may be indexed about a common centre, to be machined by a greater number of operations than would otherwise be possible. The box chuck can be indexed through 360 deg. about a vertical axis.

Box Tool See **Boring Tools.**

Box Turning Tools See **Tool Shapes.**

Brass Grinding Wheels Wheels of brass occasionally adopted in **electrochemical grinding** (q.v.) as a means of cutting cross-sections less than 0·015 in. thick or where dielectric circulation is difficult.

Brass Tools Tools used in **ultrasonic machining** (q.v.) for cutting glass and steatite.

Brazed Carbide Tools See **Carbide Tools.**

Bridge Reamers Reamers (q.v.) designed to finish holes in steel plate and parts of constructional steel. Their ends taper to enable the whole reamer to pass into a hole and bring it to the desired diameter by enlarging it. They are not designed for producing holes that taper, and are made with straight or helical flutes as required, their diameters ranging from $\frac{13}{32}$ in. to $1\frac{1}{2}$ in.

BROACHES

Broaches Tools for cutting metal, comprising a plain pilot portion, a tapered toothed portion and a plain shank engaging with the pulling head on a broaching machine. The teeth have normally single cutting edges, and on circular broaches run in series round the tapered portion, gradually increasing in cross-section as they near the shank. They are either pushed or pulled mechanically through a previously cored or drilled round, square, rectangular or hexagonal hole to produce the correct form. They will also finish flat surfaces, irregularly-formed holes and open grooves such as keyways in hubs. Another application is to the forming of teeth in internal gears and ratchets, splines in shafts, etc.

They are usually made of a tough, hard, sound, non-distorting steel of alloy, high carbon and case-hardening types, the precise steel depending on the service life required, the material to be cut, and the length and form of the broached areas. Relatively few are made of case-carburizing steel, and hardened high speed steel is now the most popular material for the cutting teeth of solid and some other broaches. The tools have to be heat-treated and ground, and the heat treatment is extremely important.

Broaches for finishing internal holes differ from those for finishing the external surface of a component. Those to be pushed through the work are shorter than those to be pulled through. For internal work the tools are usually solid, unless their diameter is so great as to render distortion in heat-treatment probable and serious. The teeth are of size and form for the operation, and if the broach is more than $3\frac{1}{2}$ in. diameter it is often made as a composite consisting of a tough central portion of carbon steel, conically ground, on to which cutting segments of nickel or high speed steel are mounted in the form of toothed rings set one over the other until the cutting portion is adequate for the job. This makes it possible to remove a segment or ring when dulled or worn and insert a replacement. Each ring then moves up one. This costs far less than replacing the entire broach.

In intricate small broaches or those of expensive type the final cutting teeth only, numbering four or six, may be made on a number of elements, that part of the broach receiving them having no teeth. This again prevents premature scrapping of a costly tool because a few teeth have lost their cutting efficiency.

The number of teeth in a broach depends on the amount of metal to be removed and the cut depth allowable for each tooth. Push broaches are mainly applied to bringing holes in bushes to precise dimensions. A series of broaches may be needed for some operations, one succeeding the other, each bringing the hole close to the final size. This is necessary, however, only when one broach alone cannot remove all the material or the hole is of great depth. The overall length of a broach includes the shank and the plain or pilot portion at the finishing end.

The teeth of external broaches are little different from those of internal type, but more than three sides cannot be used for cutting, as one side has to be used as a support to minimize chatter. Many composite external broaches have their elements inserted in a body of good carbon steel, each tooth being separated from its neighbour by part of the solid steel body. These composite broaches are not suitable for surfaces of inadequate depth. Keyway broaches are sometimes tipped with tungsten carbide brazed on to the bodies of the teeth, but these cannot be used for heavy roughing work, as this causes chatter and cracks or fractures the carbide tips. They are, however, excellent for finishing.

Round broaches sometimes have renewable finishing teeth, or their teeth may be helically cut like a screw thread, but more usually they are circular with parallel edges. To produce rectangular holes, the broach teeth may be staggered. In normal round broaches the pitch or spacing of the teeth may be either equal or variable, the latter being adopted to minimize chatter, but it must always produce chip clearance.

There are broaches for forming helical grooves, rifling gun barrels and burnishing holes previously drilled, bored or roughly broached. **Burnishing broaches** are also used to rectify hollow bushes or other cylindrical parts whose holes or bores need restoration to their true shape and size after distortion. Such broaches are built up of polished rings of round cross-section, but their teeth have no cutting angles. They are pushed into the work, and the teeth compress the internal surface, removing irregularities and giving a characteristic surface finish. They are mostly used for non-ferrous metals, and have rings of solid tungsten carbide. (See also **Broaching Machines**.)

Broaching Machines Broaching machines force special tools of tapered form, known as **Broaches** (q.v.) through a hole or over a piece of work. As the broach is pulled or pushed through or over the work, the teeth cut successively, beginning with the smaller teeth, which enter the hole, and finishing with the larger teeth, which bring the hole to finished dimensions. The intermediate teeth remove most of the metal. Broaching is mainly used for machining out holes or other internal surfaces, but is also used for flat or external surfaces, and for burnishing previously formed holes.

Broaching is particularly adaptable to producing large numbers of identical parts required in many instances to be of specific sizes within exceptionally close tolerances. Typical examples are the

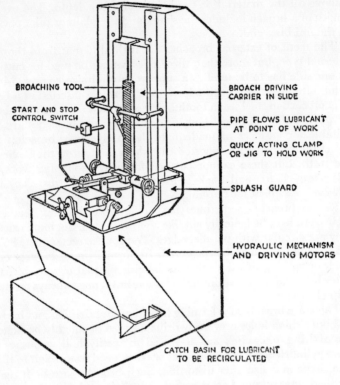

Fig. 4 Surface broaching machine (vertical)

forming of hexagonal or square holes from round, the cutting of keyways and the production of symmetrical or asymmetrical holes, grooves or slots. Other internal broaching work is forming splines or longitudinal keyways in shafts, internal gears and the internal surface of a component, as well as the completion of square holes. External operations cover forming the external surface of a workpiece, as in making the wings of shock absorbers, and are often termed surface broaching. The work must be massive enough to withstand the severe pressures involved and be firmly secured to the work table.

Broaching does some operations better than reaming, as the teeth of the broaching tool keep their shape and size longer and give greater accuracy. The machine itself may be either horizontal or vertical, and may either push or pull through the work. Broadly

BROACHING MACHINES

speaking, pulling is done by horizontal machines, but vertical machines can carry out either operation, while some of these draw the broach up through the work, others drawing it down. The horizontal broaching machines are mainly devoted to internal work, but some also do external work, whereas the vertical machines are mostly used for outside and flat operations. A type of broaching machine also exists which draws the work over a fixed tool.

Pull broaching is largely used for internal work, which can, however, also be done by push, but the push needs shorter and stiffer broaches, so that there is a limitation on its applicability. The more severe stresses set up by push broaching would bend or break the tool if it were of the same length and rigidity as in the case of pull broaching, and the work would consequently be less accurate. On occasion a progressive series of broaches is pushed through the work to give adequate precision of finish, these broaches being of increasing dimensions.

The normal horizontal pull-broaching machine is made up of a work-holder and a means of supplying coolant to the broach, which is secured to a drawbed, itself secured to a large screw working in a nut. Movement of the nut in either direction gives forward or backward motion of the screw, so giving the required pull or taking the tool back to its first position. Pull length is controlled by adjustable tappets (small pins or studs) which stop the movement at the required point. Some broaching machines will carry and work more than one broach simultaneously or alternately.

A different machine replaces the drawhead by a rack and pinion gear along which a toothed pinion wheel with a fixed centre travels to give the reciprocating motion. Push broaching, on the other hand, uses a vertical press, but is limited in its range of operations. For short runs on light work, a hand press may be employed if desired. Most push broaching machines are actuated hydraulically, rather than pneumatically, or mechanically by means of a motor-driven pump acting on pistons within a cylinder. Hydraulic operation is effective, controllable in speed, has low maintenance cost and produces the uniform motion for close limits. Speed of cutting may be 20–30 ft./min. The return stroke is much faster, to economize in time.

Broaching machines include the double slide in which one slide goes down as the other rises so that the work is continuous, the work-tables being often synchronized to suit the slide motions. Automatic broach manipulation draws the broach *upwards* and is employed for enlarging holes, forming bushes, splined holes and small internal gears. The broaches may be carried by a special handling slide and their upper ends guided until the finishing teeth approach the work. Automatic liberation of the upper broach holders from the handling slide then occurs. The tools are raised and automatically re-secured to the special handling slide, which elevates them until they have made room for the work-table to be reloaded. The broaches used in this machine have a shank at each end for holding purposes.

A special broaching lathe is sometimes used for the outer surfaces of cylindrical parts, and in this the work rotates on its axis throughout the stroke. The machine slide, hydraulically operated, traverses the tools tangentially past it, roughing and finishing in a single stroke. There is also a continuous broaching machine with fixed tools and a range of workholding fixtures fastened to a continuous chain. The broaches are housed in a horizontal tube through which the parts move towards them, bolting down and discharge being automatic. The machine gives continuous output and is used for flat and irregular surfaces that can be tooled with a direct forward motion.

Broaching Oils Oils employed to eliminate adhesion to the work broach of the chips produced in broaching. They ensure the best finish, accuracy to size and service life of the tool, if other requirements have been met. Typical broaching oils include sulphurized oils (q.v.) and special high fat-content oils, but water-soluble oils (1 part oil, 15–20 parts water) are suitable for steel, and light oils for aluminium. Cast iron is often cut without an oil. The oils cost more than ordinary machining oils and may discolour some of the metals broached. They must be completely eliminated from heat-resisting alloys in advance of heat treatment or service. In some instances they are less effective than soluble oil emulsions, being more viscous and therefore less swift in cooling and eliminating chips. Copper alloys that have to be free from stain or discoloration can be broached in paraffin containing 10–20 per cent lard oil.

Buffers Oils and fluids introduced into **honing** lubricants to reduce or eliminate chatter of the honing stones, so prolonging their life by taking up shock and recoil from variations in power. The most suitable are animal oils such as lanolin, lard oil and tallow, but proprietary oils of special compositions are also extensively sold. These incorporate rust preventives and deodorants, being diluted with about 95 per cent paraffin. The buffers should be filtered and generously used. Water and unwanted oil should be excluded. The working temperature of the oils should not exceed 17–20 deg. C. (62–68 deg. F.).

Buffing The operation of polishing metallic materials with a buff or rotating disc made up of layers of cloth charged with a powdered abrasive.

Burnishing Commonly a finishing operation designed to produce a superior surface to that given by broaching (q.v.), a higher degree of precision to the diameter of a hole and a superior wearing surface. It is performed with special broaching tools, one of which does burnishing only, while another both cuts and burnishes, and is furnished with buttons following on the burnishing teeth, their action being only that of smoothing and cold working.

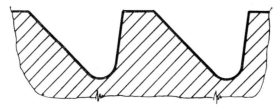

Fig. 5   Buttress thread

Buttress Thread A form of screw thread taking a bearing on the straight side only, and extremely strong in this direction. See Fig. 5.

C

Caliper Gauge A type of gauge used in measuring parts to ensure that they are of the right dimensions and form, both inside and outside. See **Gauges**. The micrometer caliper gauge is made up of a screw whose barrel is divided into small parts enabling it to measure amounts of revolution up to 1/10,000th in. Some caliper gauges of this type have their jaws square with the blade and are adjustable by vernier (q.v.) micrometer.

Carbide-oxide-and-metal Tool Material A new proprietary tool material which combines the properties of ordinary ceramics and finishing-quality carbides. A composite of oxide, carbide and metal with a wear-resistant structure, it cuts at high speeds, has good chip and heat-shock resistance, and is used for finish-machining cast iron, hard alloy steels, case-carburized steel and low carbon steels. In tool form it carries out finish- and semi-finish-turning, boring, facing and broad-nosed work, but if interrupted cuts are required, negative rake is given. It is claimed to give 300 per cent higher production with 3 times as many pieces machined as carbide or ceramic tools.

Carbide Tools Often known as **cemented carbide tools**. They are moulded and sintered products of a suitable form of carbide. **Tungsten carbide** (q.v.) is the most used, but **titanium** and **tantalum carbides** (q.v.) have also been used to a lesser degree. The metallic material is powdered and then moulded into the form of a tip, after which it is sintered (heated to an extremely high temperature), and becomes exceptionally hard. The metal from which the powder is made is highly heat-resistant and of considerable fineness, hardness and compressive strength. Small contents of other carbides are sometimes introduced, and among these are the carbides of niobium (colombium), molybdenum, vanadium, chromium, zirconium and hafnium. In addition to these an essential ingredient is a bonding agent to bind the particles together. Cobalt is mostly used, but in a few instances nickel and iron have been used.

The two principal classes of carbide tools are those containing tungsten carbide, designed for the machining of cast iron, austenitic stainless and other steels, non-ferrous alloys and other difficult metals; and secondly, those containing in addition to tungsten carbide suitable amounts of either titanium carbide or tantalum carbide. These are designed for the machining of ferritic steels. All qualities embody the necessary cobalt content for binding purposes.

The resultant metals are extremely hard, and once sintering has been carried out, all cutting edge form is achieved by grinding. The tool tips produced are brazed on to a low carbon steel shank and given the requisite tool forms by either the manufacturer or the user, suitable blank carbide tips being supplied for the purpose. In most instances, however, standard tip forms are stocked and supplied as ordered from catalogues.

So tipped, the tools will machine at much higher cutting speeds even than tools made from the best possible high speed steels, and will moreover have a much longer service life. They are widely used for mass production machining when a light cut at high speeds is more important than a heavy cut. The tungsten carbide tools are supplied in a range of standard and special grades from which that suitable for the specific operation can be chosen. Each grade has qualities suiting it to its work. Grinding of the tips is performed with silicon carbide or diamond abrasive wheels. The tools are not so strong as solid high speed steel tools, and must not be fed into the work heavily; but properly fed, they will give by and large about five times as great a removal of metal in a specified time as a high speed steel tool. They can be safely used for machining cast iron, non-ferrous alloys, hard rubber, fibre, slate, marble and different qualities of steel, and are specially suitable for machining austenitic (14 per cent) manganese steel.

In addition to lathe tools, the carbides are used for tipping twist drills, milling cutters, reamers, planing and shaping tools, and also for the tips of circular wood saws. Typical trade names are Allenite, Ardaloy, Carboloy, Widia, Wimet, etc.

See also **Cermet Tools.**

Carbon Dioxide A colourless gas generated by burning carbon in air, as when magnesite and limestone are combusted, or by other processes. It has been advocated as a coolant for certain machining operations, such as drilling steel sheets with carbide-tipped twist drills, but in at least one instance there was no advantage over dry drilling.

Carborundum See **Silicon Carbide.**

Cast Electrodes A type of electrodes used in **electrical discharge machining** (q.v.). They may be produced by die-casting from typical zinc base die casting alloys or from alloys containing 80:20 or 50:50 zinc-tin. Aluminium alloys may also be used, and it is possible to melt up the alloys and use them again. The aluminium

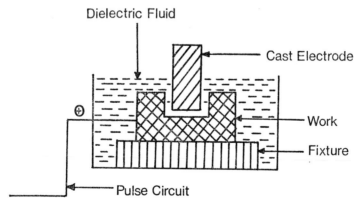

Fig. 6   Electrical discharge machining

alloy electrodes are less easy to manufacture owing to their higher melting temperature and contraction rate (5–7 per cent), but both types of alloys are sufficiently ductile to be produced to extremely precise dimensions. The electrodes are furnished with holes for the passage of the dielectric fluid, these being of the maximum possible diameter so that the fluid can pass at low pressure with a high rate of flow and enable roughing cuts to be taken. (See Fig. 6.)

Cemented Carbide Tools See **Carbide Tools.**

Centre Location The measuring of the distances of hole centres from each other, especially when the holes are not specifically related to one another, but are precisely located in reference to a common point, such as the centre of a different hole. It is necessary to consider both the angles and the centre dimensions. The necessity for the measurement often arises in the location of jig bushes and die holes.

Centre Punch or **Centre Pop** A punching tool having a cone-shaped point to indicate by a small indentation of the work the centres of holes to be drilled.

Centreless Grinding A method of grinding unmounted metallic components whereby the piece is supported by a work-rest and caused to travel between an abrasive grinding wheel revolving at high velocity and a regulating wheel revolving at low velocity. This second wheel rotates in the direction opposite to that of the first abrasive wheel. The operation is designed for cylindrical work such as solid bars, pins, rings or tubes. Embodied with the work-rests are guides to direct the work towards the two wheels and take it away as soon as the work is done. The grinding wheel forces the work by its pressure to make contact with both the regulating wheel and the work-rest.

The two primary centreless grinding methods are through-feed and in-feed grinding. In the first, the workpiece travels sideways or axially between the wheels, so that this method is confined to those cylindrical shapes without interference from shoulders. In the second method, the work does not move axially, but has a shoulder or other projection larger than the diameter to be ground, is tapered, or has a complicated profile.

A point on the work surface in contact with the abrasive wheel rotates identically with that wheel as regards direction, but at a lower speed. In consequence the surface produced is closer to dimensions. The speed is chosen to suit the work, which is rotated at a speed identical with that of the regulating wheel.

One other method of grinding designed solely for work having a taper form and front-fed by hand or machine to a fixed end-stop is

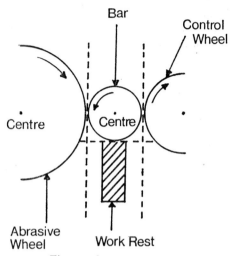

Fig. 7   *Centreless grinding*

termed end-feed. The abrasive wheel, regulating wheel and work-rest, have a fixed relation to each other, while the abrasive wheel and its partner have a taper corresponding to the desired form.

Centreless Lapping This is carried out on machines largely akin to those employed for **centreless grinding** (q.v.), the principle being the same. The lapping machine, however, gives much more accurate finishes (diametral accuracy 50 millionths of an inch, roundness 25 millionths). The wheels used are wider than for centreless grinding and are made of a bonded abrasive. Both abrasive and regulating wheels have axes out of parallel, the lapping wheel having a negative angle of −4 deg., the abrasive wheel a positive angle of 1–3 deg. Lapping marks are prevented, but lapping is done in three separate stages using each time a lapping wheel of increasing fineness supplied with a paraffin or other clean lubricant. Stage 1 is the resting of the work on a steel- or carbide-faced work rest blade, while the work-centre is a little above the centres of the wheels. This is designed to put right any lack of circularity. Stage 1 eliminates not more than 0·0005 in. of material with a finish of 4–6 micro-in. Stages 2 and 3 use a rubber work-rest, the work centring on the two wheels to prevent scratches as far as possible, while not more than 0·0001 in. of material is eliminated, with a finish of 2–3 micro-in. in Stage 2, Stage 3 removing virtually no metal, but giving a finish of about 2 micro-in.

The operation is capable of giving a high rate of output, using continuous, automatic or hand feeding on parts ranging in length from 15 in. to 15 ft. long and ¾–6 in. dia., such as shafts, bearing races, piston pins, etc. The work must, however, have been ground to the desired degree of circularity and straightness before the operation is attempted.

Centreless Roll Lapping A method of centreless lapping designed for extremely small quantities, mostly below 10. Its principal advantage is speedy setting up of the work, and for this reason it is easily applied to numerous changes of size in small quantity production. The amount of material removed is normally from 0·0002–0·0003 in., according to the finish originally given. The work is passed between a pair of cast iron rolls, 6 in. and 3 in. dia. respectively, then forced down by a grooved stick having a vee angle of 120 deg. Abrasive is applied in the form of a compound to the rolls, which rotate identically, but in the opposite direction to the work, the 6 in. roll at about 180 r.p.m., the 3 in. roll at about 90 r.p.m. The stick reciprocates to less than ½ in. from either end of the work. Maximum surface finish and dimensional accuracy are obtained by slow reciprocation.

Centreless Thread Grinding A method of grinding screw threads on components that have considerable variation in form and dimensions, and where a high rate of output is desired. It employs wheels having either a single rib or a series of ribs. The grain size of the wheel is governed by the threads per inch, i.e. from 80 to 280. The wheels used are much harder than those used in the previously described centreless grinding operations. The service life of the wheel is controlled by the hardness of the workpiece material, the thread limits and the output rate required. Regular wheel dressing is essential, the method adopted being **crush dressing** (q.v.). If high precision is required, the regulating wheels have to have a much finer grain size. The normal procedure is for these wheels to rotate identically with the abrasive wheel as regards direction; but if the threads to be ground are from, say, 1–8, the regulating wheel revolves counter to the abrasive wheel. The first type of revolution is termed down-grinding, the other up-grinding. Dressing is done with diamond tools.

Threads can be ground as fine as 400/in. by special methods, but the general run of work rarely exceeds 80/in. and is usually less. Hard threads are obtainable because the work can be ground in the fully hardened condition, so that any distortion resulting from heat treatment of the material can be ground away. Centreless thread grinding is suitable for almost every material and degree of hardness.

Ceramic Tools One of the most recent developments in cutting-tools. They are not made of tungsten carbide or metals such as steel, but of alumina with additions of other ceramic materials introduced as a means of refining the grain structure or to give better sintering. They are usually manufactured and supplied in the form of tips, sintered or pressed while hot to suitable tool forms and brazed, or alternatively soldered, to low carbon steel shanks by special techniques. The brazing and soldering methods adopted for tungsten carbide tips are not suitable, since it is difficult to obtain satisfactory "wetting" between the ceramic and the metal. The actual solder or brazing medium used is capable of accepting the varying heat expansion of the two substances, ceramic and steel, but at the same time remains sufficiently strong to withstand the stresses of machining.

Ceramic tools may also be used in the form of inserts, i.e. pieces inserted at the working faces of tool bodies. They are extremely

TABLE I
Speeds and Feeds of Ceramic Tools

Material	Speed (s.f.p.m.)	Feed (in./rev.)	Cut depth (in.)
Steel 170–207 Brinell	970	0·013	0·005
Grey Iron	210–1415 (min.) (max.)	0·010	0·010
Non-ferrous metals (excluding titanium, aluminium, brass, molybdenum)	60–3,000 (min.) (max.)	According to operation and material	According to operation and material
Plastics	300–3,000	,,	,,
Rubber	1,000 (max.)	,,	,,

hard, chemically inert and capable of considerable resistance to wear. These properties have led to their being adopted for certain types of machining, such as on cast iron and hard steel, when they will give fantastic cutting speeds if properly used and applied. See Table I. The chemical inertia of the materials means that chips do not readily weld to the tool nose owing to the heat generated in cutting, and is also beneficial in giving an excellent surface finish.

Broadly, the ceramic tools can be applied to the general machining of steel as long as continuous cutting can be ensured and negative rakes adopted. The greatest drawback of these remarkable tools is their liability to brittle-fracture, often occasioned by fatigue. Otherwise they have greater wear-resistance than, for example, **carbide tools** (q.v.), but against this, tool wear is sometimes experienced when there is chemical affinity between tool and work, as when magnesium at certain temperatures (e.g. 900 deg. C. = c. 1,600 deg. F.) causes rapid wear of the aluminium oxide. There are other examples.

In general these tools are not suitable for machining aluminium alloys, titanium and zirconium. For the best results they must be run at high speeds on the materials for which they are suitable, as otherwise they are liable to snip or fracture. Another important point is proper choice of tool form to give maximum cutting power. Deep cuts with light feeds are more effective than light cuts at heavy feeds. The ceramic inserts are about twice as expensive as inserts of tungsten carbide, precision-ground.

Ceramic tools have largely been used for turning operations at high speeds on steels and cast irons, and on non-ferrous and non-metallic materials. They perform satisfactorily on hard cobalt base alloys, many hard stainless steels, but for soft, work-hardening stainless steels and nickel alloys they are not recommended because of their innate lack of resistance to shear and their largely negative rake angles. On the other hand, in some operations they will give a cutting speed almost double that of a carbide tool on the same material and operation. They are also effective on plastics and rubber.

Cerium Oxide A medium-hard abrasive material having a grit size of 1 or 2 microns, used for lapping when a high degree of polish is required.

Cermet Tools Tools for cutting made of titanium carbide or chromium carbide (70–80 per cent) to which molybdenum carbide is added, nickel being introduced as a binding medium. They differ from the tungsten carbides in needing a higher sintering temperature owing to the nickel content, with the result that they are less tough and can be used with effect for light finishing cuts only unless a high proportion of fractures can be tolerated. It is claimed that given suitable operations and conditions they show a considerably higher rate of output than tungsten carbide. In addition they have considerable resistance to oxidation, high hardness, resistance to thermal shock, comparatively low density and good creep rupture properties at temperatures between 980 and 1,200 deg. C. (1,800–2,200 deg. F.). They have a much lower impact resistance than the carbide tools.

Principally they are used for cutting steel and cast iron in finishing operations at high speed with medium to light chip loads. They may be mechanically secured to a toolholder or brazed to a low carbon steel shank by special techniques.

Chain Broaching Machines Machines for broaching in which an endless chain is mounted on sprocket wheels and carries fixtures into which workpieces are placed and held for broaching. They are then precision-guided through a fixed number of broaching tools which carry out the work. Both loading and unloading may be done automatically or by hand. The machines are employed on a wide range of forms and for a high rate of output, but their use is restricted to some extent by the form of the work. The machines themselves are horizontal and continuous in operation, but only if the workpiece has one open side. See **Broaching Machines.**

Chamfer A sharp edge given to a circular or rectangular metallic or other plate by bevelling.

Chamfering A machining technique designed to give a taper to plates, threads and gears. It may be done in the lathe, by multiple operation machining, or combined with boring, broaching or drilling. Broach teeth often have their sides chamfered to lengthen their life, and this practice becomes increasingly advantageous as the material on which they are required to work becomes more difficult to machine.

Chasers See **Threading Dies.**

Chaser Taps Tools for tapping (q.v.) consisting of a body into which four or more chaser taps are inserted and secured by means of wedges, screws or grooves, or by screws and serrations cut into the chaser body. The chasers are often narrower at the cutting end than at the base, and are stopped by a shoulder on their body, with hardened and ground wedges to keep them firmly in position. They have carefully-cut threads and are high in first cost, but are extremely economical as compared to solid taps when a high production rate is required.

Chatter Vibration at the nose of a tool caused by slackness in the machine, lack of rigidity in the toolholder, improper design of the tool, loose bearings, or inadequate design of the machine tool causing

CHEMICAL CONTOUR MACHINING

it to be unable to function at the speed of which the cutting tool is capable. Chatter not only produces an unsatisfactory machined surface, but also gradually weakens the cutting tool by fatigue, so that it fractures or chips too soon at the cutting edges.

Care must be taken, therefore, to ensure that modern cutting tools of tungsten carbide, ceramics, super-high-speed steel, Stellite, etc. are not used in out-of-date or ill-maintained machine tools.

Broaching tools are sometimes given helical teeth to prevent chatter when broaching round holes. Alternatively, for flat surfaces or a number of internal splines about a circumference, the broach teeth may be staggered lengthwise to give a more even cut. Such broaches cost more than the standard forms, but often pay for themselves many times over by the elimination of chatter and breakage.

Rigidity and absence of chatter are essential in **milling** (q.v.) as otherwise heavy tool-wear and fracture, inaccuracy to size and bad surface finish may result. Settings are often made more rigid by using stiffer arbours, giving additional arbour support, or extra support between gang cutters (q.v.). Fixtures may also be made more rigid.

Chatter in grinding operations is liable to cause **chatter marks**, which mar the surface of the work. They are the consequence of poor balance in a wheel during the grinding of a surface, or excessive wheel hardness; inadequate lubrication, especially of the work centres; improper alignment of the centres; bad machine conditions, such as badly or incorrectly spliced belts, loose bearings, slipping, worn keyways or gearing, poor foundations; or vibration of the type mentioned earlier.

Chemical contour machining, sometimes termed **chemical milling,** differs from chemical blanking in that it aims to produce three-dimensional forms from comparatively extensive surfaces. In most instances the operator desires to form shallow recesses or irregularly-formed cavities, lessen weight, give work a tapered form, cut down the dimensions or eliminate an objectionable surface later. Its primary application is to the machining of complicated shapes and cross-sections of considerable size either too hard or lacking in ductility to be dealt with by ordinary machining without fracture, or too thin to withstand the stresses imposed by mechanical working.

Delicate work can be chemically machined after heat-treatment, so preventing warping and distortion caused by heat and quenching. The process has proved extremely useful in the manufacture of parts for astro-vehicles, etc.

The work is first degreased or cleansed in a solvent or, if preferred, cleaned in a bath of alkali. Alternatively it can be pickled in a pickling solution. It is then etched by spray or immersion, using a suitable reagent, all disfiguring dirt being removed by desmutting or bright dipping. This is the procedure adopted for the elimination of metal from all parts of the work surface.

Where, however, parts of the surface are to be left unattacked, the work is degreased and cleaned as for the previous operation. It may then be given a conversion coat, after which the masking material is applied, dried and cured, and the necessary pattern marked with a scribing tool through the maskant, which is then eliminated from the parts to be attacked. The etch is applied by spray or immersion, followed by desmutting or bright dipping. The rest of the masking material is removed, and the work is then complete.

The advantages of the process are that simultaneous etching of a large number of parts can be carried out in a single container and material can be etched away from a single side or a number of sides at one and the same time. The dimensions of the work depend on those of the tank, and may be up to 7 ft. wide × 50 ft. long. Complex forms and thin cross-sections can be made to fine tolerances, and hard or inductile metals produced commercially. The design is not restricted by form or tool conditions. Upkeep and tool costs are small and service life long. Production follows within a few hours of design, which can itself be altered simply and cheaply by the production of a different template. No rough edges are produced, and many forms incapable of being made by other mechanical processes can be obtained in this way. Distortion is also avoided.

On the other hand, metal is removed slowly (0·001–0·002 in./min.). The operation does not lend itself to long runs. The cut depth does not exceed $\frac{1}{2}$ in. No sharp inside radii can be produced. Internal surfaces of bores or cylinders where welding or brazing has been carried out, or where blowholes or other surface defects occur, are unsuitable for chemical contour machining, as are clad sheets, except in special circumstances. The material of the work must be uniform in structure and composition. Surface defects already existing in the work material can rarely be modified or removed. Safety precautions are essential.

Chemical Machining A method of producing suitable forms and sizes of parts by the removal of metal from their surfaces after chemical action. This may take the form of simple etching or a general elimination of metal under carefully regulated conditions by chemical attack. The elimination may be selective instead of general, if desired, in which case the area not to be attacked is shielded by suitable maskants (q.v.). There is little restriction on the metals capable of being machined in this way, and the tech-

CHEMICAL MACHINING

niques involved correspond largely to those used in normal etching for microscopic work, etc., giving a more brilliant appearance to a metallic surface by chemical methods, such as polishing by the aid of chemicals, and etching as applied to machine instruction plates, half-tone and zinc blocks for printing, etc. The acids and other reagents used are broadly the same, as are the working precautions and processes.

The techniques are divisible into two main groups: chemical blanking, in which parts are produced from sheet metal, and chemical contouring or milling, in which metal is eaten away from thicker sections or sheets. Chemical machining is applicable to work that cannot be dealt with by ordinary machining operations or for which these would prove too costly or difficult owing to the tenacity, lack of ductility or hardness of the metal, or because the working material would be too thin when finished, too intricate in form, or unsuitable in dimensions, for conventional machining.

Chemical blanking is applied primarily to thin sheet and foil. The work is first cleaned by immersion in a solvent, degreasing or chemical cleaning. It is then rinsed or sprayed with solvent, pickled in an acid or alkaline solution, cleaned by a steam or other form of blast, or dipped in acid, dried, and masked usually by a photosensitive process to show the areas not to be attacked by the acid. The masking is applied by dipping, whirl-coating or spraying, according to whether uniform thickness, versatility or simplicity is most required. The work is then dried at room temperature and cured by baking for about fifteen minutes at not more than 120 deg. C. (250 deg. F.) to eliminate all traces of solvent.

Meantime, the master (q.v.) has been prepared by oversize artwork, photographically reduced in size, and multiple image masters prepared from it. The workpiece is exposed to ultra-violet light between a pair of masters, and the work then developed by immersion in organic solvent or spray developer, dried, and heated to give it greater toughness and resistance, at the same temperature and for the same period of time as for baking, which eliminates traces of developer and gives the masking greater hardness and increased resistance to the chemical attack.

The unmasked metal is now immersed in or sprayed with the reagent, after which the masking is scrubbed off or dissolved by chemical solvent spray or immersion in a bath of acid.

This done, the work should be complete.

There are of course alternative methods of masking, and not all the stages of work preparation outlined above are essential in every instance. Sometimes a stage or stages may be left out.

CHEMICAL MILLING

Chemical blanking is particularly suitable for making complicated, thin, delicate parts not suitable for press work, and where freedom from rough edges is desirable. The quantities are usually in the region of 5,000 if economy is to be achieved. Parts that have been made in this way include laminations for electrical work, discs, gaskets, rotors, meter parts, templates, etc. Television parts have also been extensively made. Wherever the material is too hard or poor in ductility to be blanked out mechanically, chemical blanking is simple and advantageous. The process is also useful for minimizing assembly time and preventing rough edges.

The hardness of the work material does not affect the facility with which the process can be carried out, nor impair the various characteristics of the material. It enables designers to achieve a considerable degree of flexibility; parts are turned out not long after they have been designed; costs are low and changes in design can be readily made. On the other hand, considerable skill is necessary for the best results; the user must guard against corrosion by the reagents and their vapours and protect other workers from their effects; the outputs achieved are not large; and there is a limit to the thickness of metal that can be blanked, namely about $\frac{1}{16}$ in. It is not possible to produce sharp radii.

Chemical Milling See **Chemical Machining.**

Cherry Cutter An American term for a type of milling cutter (q.v.) designed for finishing the interior of a die or comparable tool. It is a reaming (q.v.) rather than a milling tool.

Chip-breakers In the machining of metals, especially steels, the chips produced by cutting collect on the tool nose, and by virtue of the great frictional heat generated, are liable to weld themselves on to the surface of the tool. Thus they create an obstacle to the smooth, steady flow of chips across the tool face. This may cause damage to the surface of the work as the chips accumulate and are unable to break free. In consequence, it is customary to provide a circular cavity or recess immediately behind the tool nose. This not only prevents the red-hot chips from flying off and causing injury to operators, but also reduces the frictional heat at the tool nose and immediately behind it. When tungsten carbide tools are used, the most usual procedure is to grind a half-channel longitudinally along the tool nose, but this is not taken right to the nose, its distance from the cutting-edge being greater in proportion as feed and work hardness increase.

An alternative type of chip-breaker is independent of the tool, being held on top of it and firmly secured in place by dogs (q.v.). The advantage of this is that no costly steel is wasted. It consists in

CHIP CONTROL

most instances of a rectangular piece of steel of the same length as the tool, one end having a slight concavity or convexity.

Chip Control In some internal machining operations, particularly boring, control of the flow of chips away from the tool nose is of considerable importance. The chips have to travel smoothly from the machined surface in the direction of the bore centre because, if they flow instead towards the side of the bore, they may spoil the finished surface or wind themselves about the boring tool, especially when a heavy cut is taken. This is often experienced when boring small diameter tubes. The remedy is usually to modify speed and feed by trial and error. In some operations speed has been reduced thereby by about 25 per cent and feed by about two-thirds of the nominal rate.

The problem is especially difficult in cutting die threads, where there is always a risk that the chips will be driven into a pocket or fail to leave the cutting edge, in which case the threads may be furrowed or the threading tool edge impaired. The smaller the die thread diameter, the greater the likelihood of crowding the chips. Similarly, the softer the metal the greater the chip congestion and adhesion to the tools. Chasers (q.v.) should have the longest possible chamfer (q.v.) as this reduces chip thickness and chip load per tooth.

When boring lead alloys, crowding chips may fuse together owing to the lower metal-fusing temperature, and will then injure both tool and work. The remedy is to give the tools proper relief and a large enough clearance for the chips to leave the tool smoothly and without crowding. An air blast aimed at the tool cutting-edge also helps to clear the chips.

Chip Curlers When machining ductile metals in the turning lathe, and particularly where a combination of strength and ductility is found, as in stainless steel treated by annealing, a chip breaker may not serve to break up the spiralling chips into small sections that clear themselves away or can be cleared. This means that the operator is faced with a long coiling chip continuous in form. It becomes necessary, therefore, to control this chip and guide it in a direction suitable for release, which may be done to some extent by modification of the rake and side-cutting angles (q.v.) of the tool. In addition control is achieved by cutting curler grooves in the tool, somewhat similar to those of the chip-breaker (q.v.).

Chipless Taps Where there is a danger of chip trouble in tapping aluminium, the difficulty may be overcome with the aid of a chipless tap or form tap. This necessitates tapping a wall thickness of the hole sufficiently to sustain the tool pressure. Such taps can be used for all but the high silicon aluminium alloys used in die casting, and can be

CHLORINATED OIL

given threads of 55–65 per cent. They have been run at double the speed of cutting taps proper.
Chlorinated Oil See **Cutting Fluids.**
Chlorinated Wax See **Cutting Fluids.**
Chromium Carbide See **Cermets.**
Chromium Oxide A medium-soft abrasive having a grit size of 1 micron, used for lapping soft metals or for the final lapping of work requiring a highly polished surface. It is less used than other abrasives for these purposes.
Chuck A device for gripping the work in a lathe or other machine tool (see Fig. 8). In a lathe the chuck is placed at the end of the

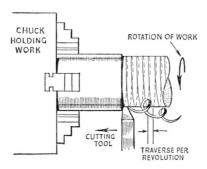

Fig. 8   Use of the chuck

spindle shaft fixed in the headstock. It has various forms, but in principle is screwed on to the spindle by a female screw fitting accurately to the screw thread cut on the spindle nose. It is more usual for the direction of traverse (q.v.) to be towards the chuck. Chucks can normally be detached and either self-centring or possessing independent jaws. The self-centring chuck is capable of centring work of simple regular shape because the jaws travel together towards the centre. There are usually three jaws. When the work is of complicated shape the chuck may have two jaws. Some self-centring chucks are provided with jaws whose teeth are threaded or curved on the back.

The so-called "universal" (q.v.) or combination chuck has three jaws which may be individually positioned at the outset, then worked concentrically. They are helpful when dealing with batches of identical components whose form is irregular.

Another chuck has four jaws and is capable of holding work of irregular shape. Each jaw can be separately moved. Chuck operation

CHUCKING MACHINES

can be pneumatic or magnetic, and in most instances chuck jaws are provided with steps and accept various diameters. They can be reversed. The pneumatically-operated chucks are often termed "air chucks". In those chucks employing magnetic attraction to hold ferrous metals, some are electromagnetic, operated by electric current, and magnetic attraction ceases when the current is switched off, so that the work can be removed without difficulty. Others employ *permanent* magnets for fairly light work of round shape.

A new power chuck of French invention has been introduced, incorporating automatic operation for gripping and slackening, with immediate application to a wide range of workpiece sizes, and precise control of external or internal gripping intensity. (British Pat. 1,144,580.)

Chucking Machines A type of machine tools automatic in operation and in essence akin to automatic turret lathes (see **Turret Lathes**). They are used for drilling, turning, boring, reaming and other work on cast, forged or other parts and bars. The essential feature is that the work is always held in a chuck (q.v.).

Chucking Reamers See **Reaming**.

Circular Form Tools See **Form Tools**.

Circular Sawing Machines Machines using a steel disc whose serrated rim is provided with teeth, set and sharpened and revolving at a high speed. Many sawing machines now have saws with teeth in the form of detachable and renewable saw segments secured firmly to a central saw disc. The work is bolted or clamped on a solid work table, normally a casting, and held in a vice of mechanical or hydraulic operation. A feed-roller is sometimes fitted and carries the work. The roller is power-operated and moves the work along into the cutting position. The saw blade is carried by a spindle running in suitable bearings, and is electrically driven, though a few belt-driven machines still exist. The saw blade is mounted on a sliding carriage moving along the ways of the bed towards the work. In other machines the work is taken to the saw by a sliding table. The saw saddle is held in position by long, narrow guides having adjustable taper gibs to take up wear and ensure square cutting.

Feed is either mechanical or hydraulic, and the saw head automatically returns to the starting position after cutting. The speed and feed can be controlled to suit the work, and an automatic mechanism prevents the saw from travelling too far either backwards or forwards. Controls are grouped so that the operator sees the saw at work. Coolant is supplied in sufficient quantity as the saw travels along its path unless the work demands dry cutting, as with brass or other light materials. A chip clearing device is incorporated.

CIRCULAR SAWING MACHINES

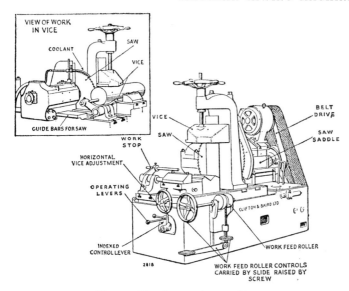

Fig. 9   Circular sawing machine

These machines will saw hot or cold steel, the tooth form being modified accordingly. Cold sawing speeds on steel are high. Solid saw blades can be run at 25–100 ft./min. with feeds 0–20 in./min. The machines have largely replaced bandsawing machines on mass-production work, e.g. the sawing of steel bars and joists. Aluminium alloys can be sawn at peripheral speeds ranging from 2,000–15,000 surface ft./min. according to the saw material, the kind of cut and the machine's ability to operate at high speeds. Feed rate on these alloys is from 4–24 in./min. according to the type of aluminium being cut. The section width and saw speed also affect the feed rate.

Of recent years saws tipped with carbide cutting blades or teeth have proved extremely successful, especially in sawing aluminium alloys, and normally give a much longer service life to the saw. The saws range in diameter from 10 to 84 in., have a number of teeth (36–190) that varies with the speed and diameter of saw, and run at a feed rate of about 0·001 in./tooth. The blade speed ranges from 480–16,000 surface ft./min.

Carbide blades are also suitable for sawing magnesium at speeds as high as 10,000 surface ft./min. A 12 in. dia. blade of this type, correctly designed, is capable of sawing 1 in. thick magnesium plate at 250 linear in./min.

Clamps Devices used for securing work firmly while being machined, ground, fitted or marked out. There are many different forms, but all have to be powerful enough to grip the part rigidly without causing it injury or undue stress, and must be swiftly worked whenever a long string of parts have to be dealt with. They must also be simple to manipulate.

Clapper Box In a **shaping machine** (q.v.) a box containing the tool, which is firmly held for the forward stroke, but is able to lift upwards clear of the cut for the return stroke after power is taken from the cutting-edge. In this way the work surface encounters only the relatively small weight of the tool, and there is consequently no damaging of the work surface. Alternatively the clapper box may be positioned at an angle out of line with the feed, and swivels away from the work surface on the return stroke.

Clearance Angles See **Tool Angles**.

Climb Hobbing A method of hobbing (q.v.) comparable to **Climb Milling** (q.v.), for finishing the sides of splines more efficiently than by ordinary methods of hobbing. It is claimed to prolong the service life of the tool by up to 30 pieces/hob, and to use less power/cu. in. of metal hobbed, while producing virtually no burring of the splines, thus eliminating a later operation. The efficiency of the operation is said to be largely independent of coolants, speeds, feeds and other factors.

Climb Milling (See **Milling**.) In normal milling with a cylindrical, toothed milling cutter, the cutting-edges travel in a direction opposite to that of the work feed. In climb milling, on the other hand, the work is fed in the same direction as the tool cutting-edge. Thus the work is being constantly forced downwards instead of upwards, as in normal milling, so that there is no risk that it will be lifted off the machine bed.

The technique has advantages and drawbacks. The cutter lasts longer and enables higher speeds and feeds to be chosen. The cutter tooth engages at the top rather than the bottom of the cut, i.e. where the chip is thickest, so that rubbing in the cut progressively declines, and the tooth edge is slower to become dull. This is particularly noticeable in cutting hard steels that are not abrasive. In normal milling rubbing is forced to occur in the cut, and speedily blunts the tool.

The downward thrust of the tool makes it easier also to clamp irregular or difficult shapes or components, the work being partly secured in position by this thrust.

On the other hand, if the milling machine is old, worn, or in a poor state, the feed screw may have some looseness or play in relation to the work-table, and then the milling cutter will pull the work in too fast, the cutter teeth cutting too deeply into the metal and being broken, or the work damaged.

Climb milling demands first-class machines and is consequently better for components whose shape renders their securing to the work table more difficult than for uncomplicated forms.

Sometimes the two methods are used in combination. A part roughly milled by climb milling may be finished by normal milling on the automatic reverse of the machine. This technique is often employed for milling chaser blanks, using two separate fixtures. When one piece is finished, it is taken away and another introduced, while milling goes on with the other.

CO_2 Coolant See **Carbon Dioxide Coolant**.

Cog-tooth Contact Wheels See **Contact Wheels**.

Cold Form Tapping This is a method of threading a part internally by displacing rather than removing metal. The form of the thread is given by a tool which has neither flutes nor cutting edges, resembling an ordinary screw as viewed from the side, but the major and minor diameters are not regularly formed, and it is these irregularities that force the material into its proper location during revolution of either work or tool. The technique is not applicable to every metal, but can be of value in tapping some stainless steels and ferrous-base powdered metal parts, as well as some of the less hard steels. It is also useful for those fibrous metals that present problems owing to the character of their chips.

Principally, however, cold form tapping is used in tapping non-ferrous alloys and the softer low carbon steels. The taps are usually of high speed steel of comparatively low carbon content. Blind holes can be readily tapped and the absence of chips is a considerable advantage here as elsewhere. Aluminium and its alloys as well as copper and its alloys can also have blind holes formed by this process.

The machine can be of any conventional type used in tapping, but the torque (q.v.) increases more rapidly than with normal tapping, the increase depending on the kind of material being tapped. Double up to quintuple torque may be required for some materials in threading, the quintuple torque being experienced particularly in machining some types of stainless metals. The tapping speed is about two and a half times as fast, but need not be high. The lubricant is a sulphurized or chlorinated oil, as otherwise the threads may be damaged. The tap drill holes need to be of greater diameter than for ordinary tapping. There is no difficulty in cold form tapping by hand.

Collapsible Taps Sometimes termed "collapsing taps", these are screw thread taps with chasers arranged to draw back as soon as the screw has been formed, enabling the tool to be removed without the need of reversing it by rotation. These taps may be either fixed or moving. The movable or revolving taps are employed in drill presses and tapping machines. The fixed taps are used in machines that cause the work to revolve, e.g. in turret lathes. The cutting-edges are automatically retracted as soon as pressure is applied to a ring or yoke arrangement. Fixed taps of this type are manually reset by lever.

The taps can be provided with either flat or blade chasers, but circular chasers can also be supplied, and are used mainly for larger-diametered work. However, if given suitable accessories, both blade and circular chasers can be inserted in a collapsible tap without change of tap. The blade chasers range in size from $1\frac{1}{4}$–4 in., while circular chasers range from $3\frac{1}{2}$–5 in. dia. Collapsible taps are perfectly designed for the same appliances as solid taps, given adequate space for the collapsing head.

The great point is that these taps can, as stated, be adjusted and taken from the piece being tapped without having to reverse the spindle. This allows the operator to modify the allowance for distortion of the workpiece material. Plated or heat-treated work can also be allowed for by adjustment to suit the plating thickness or the expansion of the heat-treated metal. In consequence, these taps can be safely adopted irrespective of the length of the run or the output required. The degree of wear arising from friction between threads and tap on reversal of the tool is minimized, as is the risk of tap fracture or thread injury resulting from chips accumulating in the hole, a particular danger with work metals that produce fibrous chips.

As against this, collapsible taps call for a more robust mechanism, need more maintenance because of the multiplicity of moving parts, are liable to produce too great a variation in the holes tapped should the parts of the tap work loose or show wear, cannot safely be employed in holes less than $1\frac{1}{4}$ in. dia., and are more expensive than solid taps.

Collet An outwardly-coned sleeve with a slot down one side, designed to be drawn into the internally coned nose of a lathe mandrel to grip a piece of work, and especially valuable when setting has to be carried out quickly and with precision. The collet has three or four jaws, and its internal opening is identical in form and size with the bar to be held. By being pressed against the internal or female cone it is closed for gripping when pulled backward, or forced forward, the better method as there is then no danger of drawing the

COMPOUND REST

work away from its stop. Closing of the collet may be mechanical or pneumatic.

Automatic feeding of the bar into the collet speeds up the operations of chucking and feeding. Bars for collets must be straight and close to diametral dimensions, the tolerance maximum being 0·005 in. to avoid excessive stress on the chucking and feeding mechanism.

Compound Rest A slide in the lathe located above a principal cross-slide below. This upper slide may be revolved to whatever angular position is required. The tool can then be presented at an angle to the workpiece. The compound rest has its base marked out in degrees whose position indicates the angle at which the upper slide is set.

Concave Milling Cutters See **Milling Cutters**.

Contact Wheels In abrasive belt grinding (q.v.), the wheels over which the belt travels. The resistance of their surfaces to the pressure of the work gives the abrasive grit of the belt surface the opportunity to do its job. The pressure of the wheels depends on their hardness, so that merely by choosing a wheel the operator automatically determines the rate at which metal is ground off, the degree of finish provided, the economy of the operation and the accuracy of the grinding. The wheels normally run at speeds from 3,500–7,500 surface ft/min., but both lower and higher speeds are perfectly attainable up to, say, 10,000 surface ft./min. The wheels are made of steel, rubber, compressed canvas, rubber-coated canvas, solid-section canvas, buff-section canvas, inflated rubber and polyurethane, and these give successively heavy, lighter, and finer grinding down to the finest polishing. The coated rubber and felt wheels are normally employed solely for special work.

Continuous Thread Rolling See **Thread Rolling**.

Contour Planing See **Planing**.

Contour Sawing A form of **Bandsawing** (q.v.) in which the sawing blade used is much narrower than the standard bandsaw, which is usually 1 in. wide or over. They are specially narrow, of alloy steel, and range from $\frac{1}{16}$ to $\frac{3}{8}$ in. wide. A special machine is required having a strong foundation and frame, as vibration must be minimized. A typical contour saw has a throat capacity of 36 in. and a saw speed range of 50–6,500 surface ft./min. according to the material being cut. The machines are capable of high output and yet are flexible and accurate enough to cope with small, intricate work. A specially important feature of the machines is an automatic butt-welding device for joining all saw widths from $\frac{1}{16}$–$\frac{5}{8}$ in., which is a great improvement on the old method of saw brazing by electrical resistance.

CONVEX MILLING CUTTERS

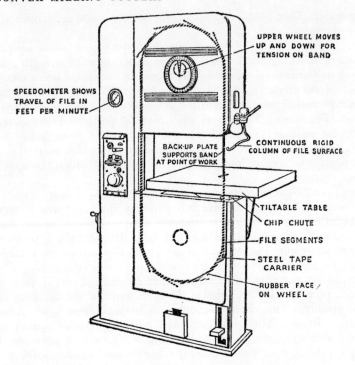

Fig. 10   Contour sawing machine

The machines may be of fixed work-table and power-table type, or of radial-arm type, the latter for large, heavy work. The maximum work height they will accept is about 55 in., and the same limit applies to thickness. Some special machines have been constructed to allow a height of up to 120 in. Fixtures and attachments are often required, such as work-squaring bars, ball transfer slips, etc. The bands are usually of carbon, high speed or alloy tool steels, but some have high speed steel teeth welded to a band of carbon steel. Others have inserts of tungsten carbide welded to the steel, but in that case a 1·3 per cent carbon steel band is mostly used, containing a small chromium content (0·2 per cent). The teeth are of standard cutting types used in sawing, but of special form for carbide inserts. There are also special saw blades for cutting difficult materials, and these may be separately classified as having spiral teeth, diamond-edge teeth and aluminium-oxide teeth.

Convex Milling Cutters See **Milling Cutters.**

Coolants See **Cutting Fluids**.

Core Drills Drills for cutting the holes made in castings by cores during the moulding process. These are preferably drilled with three- or four-fluted drills, as two-fluted drills often break when used for this work. Some of these drills have as many as six flutes. Three-fluted drills will, however, suffice for enlarging cored holes up to $\frac{1}{2}$ in. dia., and have a superior chip clearance when heavy cutting is in progress. For holes of greater diameter than this the four-fluted drill gives the greatest possible flute space combined with rigidity. Owing to their more numerous cutting edges, core drills may be employed at higher rates of feed without adding to the chip load per cutting-edge.

Corner Angles In face milling, angles equivalent to the side cutting angle of a turning tool.

Corner-rounding Cutters A type of milling-cutter (q.v.), which may be either left or right hand, used for finishing the corners and edges of parts. The cutter form is not changed when grinding the teeth faces.

Corundum A medium-soft natural alumina-type lapping abrasive for roughing, finishing and polishing. In grit sizes from 400–900 it is used for roughing and finishing the softer steels, but is not nearly so popular as silicon carbide and fused alumina.

Counterboring The enlargement of a hole for only part of its length so that a shoulder is formed at the end of the enlarged portion. The operation is often applied to holes for cap screw socket heads. The machine is identical with those for drilling (q.v.), and counterboring often follows drilling. The tools used are of two kinds, those with pilot (q.v.) and those without. The first make use of a previously-drilled hole as a means of reducing vibration and steadying the cutting action, while the second have to be provided with bushes for guidance.

Counterbore shanks normally have a long neck between shank and cutting-teeth for the boring of deep holes, the shank being either straight or having a Morse taper (q.v.). The clearance for the teeth is given to the ends only, as these do all the cutting. Clearance of about 4–5 deg. is given by forming a small neck between guide and body of the tool. Cutting speeds and feeds are slightly less than for drilling. The precise rates depend on the tool material and the work hardness, and also on the type of lubrication used. In some machines drilling and counterboring are done simultaneously.

Countersinking A machining operation in which a cone-shaped tool similar in character to a reamer (q.v.) gives a tapering or chamfered form to the beginning of a hole, which then accepts a

COUNTERSINKING

fastening device, such as a screw, whose head lies flush with the edges of the hole. The difference between this and counterboring is that the latter enlarges a hole cylindrically. The countersinking tool cuts the metal surface concentrically with the hole and at an angle less than 90 deg.

The same machines can be used for countersinking as for reaming and drilling, and in fact drilling and countersinking are frequently combined, as are countersinking and reaming. The tools may be made by grinding a twist drill to a suitable form, but it is better to use one specially made for the operation, which will have greater stiffness, and will not only give a more precise hole diameter, but also produce a superior finish. Such tools can be obtained in various diameters and cutting angles, and include machine-countersinking tools, three-flute countersinks, single-flute and insert countersinks. Of these, the first have radial relief and four flutes, and are mostly supplied with cutting angles of 60, 82, and 90 deg. for producing the bevelled edge, but other angles can be provided if required. These tools can also be used for removing burr.

The second type effectively lessens vibration and may be employed in either stationary or portable appliances, being particularly advantageous in the portable, owing to the readiness with which they centre in previously-drilled holes. They can be obtained in a wide range of diameters and cutting angles. Single-flute countersinking tools produce holes of considerably smaller diameter than the two previous types, but the hole diameter in advance of the operation is a minimum of 10 per cent of the cutting tool diameter. These tools are used when multiple-flute tools are too large for the hole diameter, and when they develop an excessive degree of vibration.

Insert countersinking tools are fastened to the body of a twist drill at the required depth so that the drill becomes a pilot and gives the tool ample stiffness. The cutting edge of the inset enters the surface and forms the previously produced hole. Such tools may also be secured and locked to taps by a spring-loaded device designed to keep the point precisely held as the tap reverses or to compensate for the coarse lead of the tap in relation to the feed required by countersinking.

In addition there are interchangeable countersinking tools for heavy work which have their own pilots; and combination tools combining the functions of countersink, drill or reamer. Countersinking speeds are usually $\frac{1}{2}$–$\frac{2}{3}$ of the twist drill speed for identical material. Feeds depend largely on the material being machined, the depth, diameter and angle of the countersinking tool. They range from 0·002 to 0·012 in. per rev. or more. Combined drilling and

CRANKSHAFT DRILLS

countersinking usually involve a slower drill speed than would be achieved by the drill working independently.

Crankshaft Drills Twist drills forming angular holes for lubrication purposes in forged steel crankshafts. They are two-fluted, one flute having a right-hand helix and the other an included point angle of 135 deg. The holes are long in proportion to diameter, so that the drills have a web of greater cross section than usual to provide adequate rigidity, but as the cutting edge of the drill becomes excessively wide, the sides have to be reduced by grinding or the drill will rub in the cut. The drills have split points and are almost always used on hard metals (up to and over 400 Brinell), the conventional drills being suitable for alloys up to 350 Brinell.

The crankshaft drills will give deep holes. On tough alloys such as nickel steels and nickel itself, they need a rather greater helix angle than usual and, if the drills are more than $\frac{3}{4}$ in. dia., grooves may be formed through the drill lip going back along the lip clearance, with staggered spaces between the cutting edges. The object is to give finer chippings which will not clog in the hole. Good cutting action is also obtained in this way, and some users thin the web at the chisel point of the drill.

These drills are also better for drilling titanium alloys than are the standard tools.

Crankshaft Lap A type of lapping tool made of abrasive cloth or paper in place of the normal abrasion stones or "sticks" of bonded abrasive material. The laps are said to give an exceptionally bright finish, with a surface as readily produced as with the more usual form of lap or abrasive.

Crocus See **Polishing.**

Cross-slide That part of a lathe or similar machine tool that provides the inward and outward or forward and backward motion making the tool cut into the work. It is a part of the carriage, travels at an angle of 90 deg. to the bed length, and is operated either manually or by power.

Some lathes have less simple forms of slide, a separate slide located above the first carrying the cutting tool. By this means the tool, pivoted horizontally, is presented to the work at any one of a series of angles, as required. A combination of this type is termed a **combination slide rest,** or **compound rest** (q.v.).

Crown Shaving A method of **shaving** (q.v.) that reduces concentration of loads at the extremities of gear teeth. Such concentration arises when the axes are wrongly aligned. The operation slightly changes the tooth contour in the radial and axial planes, and is achieved by rocking the work-table during its reciprocation if shav-

CROWNING

ing is done by the axial traverse technique. If angular traverse technique is employed, however, the shaving cutter is slightly altered for the purpose. The usual extent of crown shaving is from 0·0003–0·0005 in.

Crowning See **Shaving**.

Crush Dressing Forming abrasive wheels by rolling their working surfaces under pressure with a roll of the proper contour. This roll is most accurately made and produces the reverse of its own form on the periphery of the wheel. Wheel and roll are firmly held while in contact to ensure accuracy even when undesired deflection takes place. The roll is of hardened steel and has a diameter of 3–6 in. Crushing is usually done dry and a strong air jet blows loose grains away.

Crush dressing saves time in forming the abrasive wheel; gives a good, free-cutting surface; high output between dressings, which ensures economy and a saving in dressing tool cost, since many dressings can be given before the roll needs to be re-cut or renewed. The bond holding the grain together is broken down and releases complete grains rather than parts only. There is less risk of overheating or burning the work.

The rolls are of high speed, alloy or hard carbon steel, and are made to limits of 0·002–0·003 in. The most satisfactory results are given on aluminium oxide wheels, especially those of small grain size, medium grade and structure and vitrified bond (q.v.), when these are crush-dressed. Silicon carbide wheels can also be crush-dressed, but not resinoid, shellac and rubber-bonded wheels. Crushing pressures vary considerably according to the wheel diameter, and range from 100–150 lb./in. of wheel width up to nearly 5,000 lb. for wheels 3 in.–6 in. wide.

Wheel spindle and crushing roll must be rigid in relation to one another to prevent deflection. Wheel and crusher must run at 250–300 surface ft./min. The deeper the cut, the slower the work speed. Slower than usual work speeds are better than standard speeds. Special machines have been built for grinding threads and in these the complete crush-dressing mechanism together with a low speed wheel drive are incorporated. Existing machines can, however, be converted. In general crush dressing is for deep and intricate contours. The minimum filler and external radius produced by a crush-dressed wheel is about 0·003–0·004 in. As compared to diamond dressing, a crush-dressed wheel does not give a depth as efficiently ground as with a diamond-dressed wheel, nor is the finish as good. Wheel speed in crush-dressing is about 200 r.p.m. Ample lubricant is used.

Crush Truing A method of forming the face of a grinding wheel in which tungsten carbide, high speed steel or cast iron rolls formed to the contour required in the work are applied under pressure to the face of the abrasive wheel. The carbide rolls are used for maximum output, the high speed steel and cast iron for smaller quantities. The rolls need a spindle of robust type able to resist the side pressures. A suitable oil lubricant cleanses the wheel and lessens roll wear.

As compared to diamond truing, crush truing gives intricate contours in a single speedy operation; more sharp cutting-points on the wheel surface, so that a greater amount of material is ground off for each revolution; uniformity, form and precision of surface finish to each part ground; and greater output for each dressing, since the wheel is freer-cutting, clogging of the wheel with particles of metal being prevented even when the less hard metals are ground. It does the work at higher speed (200–300 surface ft./min.). As against this, the surface finish is inferior to that given by the diamond, the grooves are not so deep, shoulders with almost vertical sides cannot be ground; and the contour produced has less precision.

Cup Wheels Abrasive wheels given a cup-shaped form, for grinding tools such as twist drills, milling cutters, etc. They are used where the drill lands (q.v.) are thick and the cutters of diameter above 4 in. The cup or saucer wheel gives a flat land and uses a lighter cut than plain wheels, while more attention is paid to the work owing to the larger area over which wheel and work are in contact. When the cup wheel is used for grinding milling cutters, the grinding rest is set lower than the cutter, and the wheel plane at a slight angle to the cutter axis to enable the more distant face of the cup to clear the cutter. The wheel head is raised or lowered to give the required clearance. See Table II (page 54).

The cup wheel travels in a straight line along the tooth length, and its sides grind the adjacent sides of a pair of teeth simultaneously. The tooth of the gear, if gears are being ground, roll against the wheel, whose sides are chamfered to an angle equal to the pressure angle desired on the teeth. In this way a cup wheel generates teeth on spur and helical gearing up to 18 in. in pitch diameter, but tip and root reliefs cannot be generated.

Cup wheels can be satisfactorily used in surface grinding the ferritic, austenitic and martensitic types of stainless steels, and for grinding the teeth of non-generated, formed gears, producing a curved tooth with straight-profiled sides and roots, both ground simultaneously without generating motion. Complete finish-grinding of, for example, a bevel gear, is achieved with 3–4 revs. of the work, a pre-established amount of material being ground off at each

Table II
Cup Wheel Clearance Table

Wheel Dia. (in.)	D. for 5 deg. clearance for setting tooth rest when grinding peripheral teeth on milling cutters	D. for 7 deg. clearance for setting tooth rest when grinding peripheral teeth on milling cutters
$2\frac{1}{4}$	·099	·139
$2\frac{1}{2}$	·110	·154
$2\frac{3}{4}$	·121	·170
3	·132	·185
$3\frac{1}{4}$	·143	·200
$3\frac{1}{2}$	·154	·216
$3\frac{3}{4}$	·165	·231
4	·176	·246
$4\frac{1}{4}$	·189	·262
$4\frac{1}{2}$	·198	·277
$4\frac{3}{4}$	·209	·292
5	·220	·308
$5\frac{1}{4}$	·231	·324
$5\frac{1}{2}$	·242	·339
$5\frac{3}{4}$	·253	·354
6	·264	·370

Cutter Dia. (in.)		
$\frac{1}{2}$	·022	·031
$\frac{3}{4}$	·033	·046
1	·044	·062
$1\frac{1}{4}$	·055	·077
$1\frac{1}{2}$	·066	·092
$1\frac{3}{4}$	·077	·108
2	·088	·123
$2\frac{1}{2}$	·110	·154
$2\frac{3}{4}$	·121	·170
3	·132	·216
$3\frac{1}{2}$	·154	·246
4	·176	·297
$4\frac{1}{2}$	·198	·308
5	·220	·339
$5\frac{1}{2}$	·242	·370
6	·370	

CUTTER MOUNTING

rev., the wheel making line contact with the teeth. The speed used is in the range 3,800–4,500 surface ft./min., and wheel dressing is automatic at established intervals between operations. Hypoid and spiral gears can also be finished in the same manner, as well as those gears having a 0 deg. mean spiral angle.

Cutter Mounting The placing of a cutter on the spindle or arbour of a milling machine. The cutter is secured to the nose of the spindle by bolts or nut arbour, the nose face being flat, smooth and without notches or rough edges. Any part of the spindle, centring plug or arbour, making contact with the cutter rotates with perfect truth, and these parts alone touch the cutter rigidly. Driving keys, bolts, nuts, and screws in clamps must be so manipulated that they do not cause the cutter to seize up or compel the mill to run out.

Cutting Fluids Liquids employed in machining operations and directed against the cutting tool to disperse the heat generated by the friction of cutting in both tool, chips and work.

The liquids are also claimed to lessen the amount of friction between chips and tool and disperse the chips so that they do not encumber the tool in the cutting region. Cutting fluids are varied to suit the character of the operation and work, but in the main comprise mixtures of oils or emulsions.

A cutting fluid chills, and by lessening the generation of excessive heat at the tool nose prevents the softening of the tool and reduction of its cutting power by loss of "temper"; eliminates or minimizes distortion or warping of the work owing to heat; effectively lubricates both tool and work, so reducing tool wear, heat-generation and cutting-power required; gives a better surface finish; helps to give chips of the proper form and removes them, so preventing tool failure, especially when milling, drilling holes deep in relation to diameter, and hacksawing; it also prevents corrosion of machine or work by covering them with a thin film of non-corrosive fluid.

Coolants are of four main types: water solutions of alkalis, cutting oils, chemicals and solids, and emulsions.

Water is an effective coolant and inexpensive, but is unsatisfactory because it causes rust. Soluble ingredients are therefore added to prevent this: e.g. 1–2 per cent of borax, sodium resinate, sodium bicarbonate, trisodium phosphate, sodium silicate or caustic soda. These water solutions are mostly for drilling, sawing, light milling or turning.

Oils are complex and give not only lubricating power to the coolant, but also a superior surface finish to the work. They are subdivided into **fixed, mineral** and **compounded oils.** The fixed oils are of organic type, e.g. lard oil, cottonseed oil, rapeseed oil, turpen-

CUTTING FLUIDS

tine, etc. The mineral oils cost less than fixed oils and are made mostly of petroleum. Compounded oils are mixtures of organic and mineral oils, and combine the good points of both, many of them being **sulphurized** for fast cutting with a good surface finish and accuracy to dimensions on hard or difficult materials. They may have up to 4 per cent of active sulphur. Others have high contents of sulphur compounds blended with mineral oil. Both organic and mineral oils can be sulphurized, and may then be used for gear-cutting, threading, broaching and tapping carbon and alloy steels, stainless steels, etc.

Chemical and solid cutting compounds are largely of phosphorus and chlorine, graphite, mica, white lead, tallow, etc., and are not in essence fluids.

Emulsions are small particles of oil suspended in water, with alkali or soft soap sometimes added. The proportion of oil to water ranges from 1–10 oil, 99–90 water, per cent, varying with the material to be machined and the type of operation. Emulsions are not costly and may be applied to almost all machining operations.

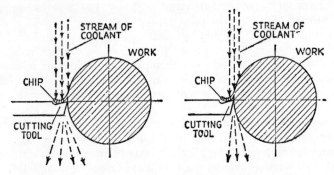

Figs. 11 and 12   Correct and incorrect use of cutting fluid

Those with the higher oil contents are for broaching, threading and gear-cutting where surface finish is not the primary objective. The lower oil content emulsions do the less difficult and demanding work, e.g. plain drilling and light turning. Typical among them are 1:4 soft soap and mineral oil, and mineral oil with neutral sulphurized oil and water. Phenol (1 gall.:1,000 gall.) is frequently added as a disinfectant.

A good average flow rate for a coolant is 4 gall./min. for each cutting tool.

CUTTING SPEEDS AND FEEDS

Table III shows the coolant to be used for a particular operation, while Figs. 11 and 12 show correct and incorrect methods of applying it with the best results. In modern machine tools coolant is reclaimed, chips being strained out and the cleansed coolant pumped back into circulation.

TABLE III

Coolants proposed for particular Operations

Operation	Coolant
Automatic Machining	Oil, compounded, mineral or organic
Boring	Emulsion (water + mineral oil + soft soap)
Broaching	Oil, compounded or mineral, or emulsion
Drilling, Milling	Emulsion as above
Parting-off	Compounded oil or emulsion
Planing	Dry
Tapping	Mineral oil + organic (lard) oil or organic oil, or emulsion as above
Threading	Compounded or organic oil or white lead + oil + sulphur
Turning	Emulsion as above
Turret Lathe work	Compounded oil or emulsion

Cutting Speeds and Feeds See **Feeds and Speeds.**

Cyclex Cutting A method of rough and finish cutting gears in a single operation used where separate machines for these two operations are not warranted by the number of gears to be machined. The gear is rough-cut and finish-cut in one chucking from a solid blank, and as the blades for finishing are located below the roughing cutters they do not come into contact with the work until the roughing cut has been completed. The work is then swiftly moved forward to enable the finishing blades to start their cut, and when this is done the gear is quickly withdrawn and indexed (q.v.), the cycle being repeated until the operation is at an end. The technique is non-generative, and can be employed for a considerable range of dimensions and for work on both gear wheels and pinion wheels.

Cylindrical Die Thread Rolling See **Thread Rolling.**

Cylindrical Grinding The finishing or truing (i.e. making the surfaces true) of cylindrical work by grinding. It is the quickest way of bringing work to the final dimension, and is more accurate than

CYLINDRICAL GRINDING MACHINES

lathe-turned work, in addition to giving a smoother surface, though lathe-turning is preferable when the primary factor is the speedy removal of metal. The technique is used for tapered parts, cams, eccentrics, shaft shoulder faces, etc., as well as cylindrical work. For high tensile steels aluminium oxide wheels are used, whereas mild steels are quite well finished with silicon carbide wheels. Vitrified wheels are mostly used, but for the finest finish a resinoid or rubber-bonded wheel may be employed.

Roughing work is done at a speed of 30–40 ft./min., and finishing 40–60 ft./min. Straight plain wheels are mostly used for simple grinding of an external surface when the wheel is capable of being traversed to give the required feed. Mostly coarse-grained wheels are used. The wheel cut for roughing is 0·001–0·004 in., and for finishing, 0·00025–0·0005 in. This, of course, is only a rough guide, and represents the depth of penetration of the wheel into the cut. Surface speed of the wheel is usually recommended by the makers for any specific type, but again as a rough guide, 5,500–6,500 surface ft./min. is advised, but up to 8,500 ft./min. can be given by high speed machines and wheels.

A coolant is always advisable. Soda and soluble oil in the proportion of 1 lb. soda and 1 pint of oil/10 galls. water is a useful mixture.

Cylindrical Grinding Machines These may be either of standard type or be "universal" grinding machines specially adapted to this particular grinding operation. They are precision machines of massive, powerful type and for preference should be designed specially for the work. A universal grinder will perform a wide range of cylindrical grinding operations with motions for each class of work, bevelled, straight, internal and external. The machines are largely automatic, but hand control for grinding small quantities can be used. Parts in quantities ranging from 75–300 are suitable for semi-automatic cylindrical plunge grinding. Where two or more external diameters have to be simultaneously ground, a profile plunge grinder is more economical, the work being done between centres. An operation of this type is restricted to parts unlikely to bend when grinding pressure is applied by the wheel, and is also dependent on there being a wide wheel and a machine having the power to drive it satisfactorily. See also **Plunge Grinding.**

Cylindrical Thread Grinding See **Thread Grinding.**

D

Damper Plugs The operation of boring a hole in metal often results in vibration and chatter, caused by the motion of the tool in the work, so that when the tool stops work the movement ceases. Some operations, however, also cause an alternating force to be developed which is independent of the motion and continues even after vibration has stopped. In other words, the vibration is self-excited, especially so when the boring tool has considerable length in proportion to its diameter. To control this type of vibration is difficult and is not wholly achieved by changing the tool material from, say, steel to tungsten carbide, though this will decrease forced vibration.

There is, however, one method sometimes adopted with success. This is the use of a **damper plug** made of a heavy material. A hollow quill (q.v.) shaft has its unsupported end bored for the insertion of the plug, which is between 0·002–0·003 in. less in diameter, the end-clearance of the plug being about the same as the clearance on the diameter. Vibration of the shaft leaves the plug motionless owing to gravitational inertia, so that the flow of air is from one side of the plug to the other. This disperses and cuts down the vibration in advance of resonance or build-up. The film of air disperses the energy that causes vibration, and the plug withstands the liability of the shaft to vibrate, much as a gyroscope withstands an attempt to change its direction. The plug causes the shaft vibrations to cease soon after its insertion. The plug has to be heavy because weight is an important factor, and a 70 per cent tungsten alloy is used, since it is heavier than even lead. At the same time excessive weight is inadvisable as it may decrease the rigidity of the shaft. The plug is best set at the free end of the hollow shaft.

The damper plug is used for shafts whose length in proportion to diameter ranges from 3·5:1 to 8:1. Plugs cease to be effective when the shaft is exceptionally long and has in consequence a low frequency. The radial clearance of the plug should be small, as the smaller the clearance the greater the allowable speed; but the speed should never be so high that the plug is drawn to one side and maintained there, which causes it to lose its vibration-damping effect.

Deburring The removal of metal from a localized area with the least possible dimensional change on the work as a whole. It is achieved by abrasive jet machining, chemical machining, countersinking, electrical discharge machining, electrochemical grinding and electrochemical machining, while it may also be combined with other machining operations.

Deep Holes Holes of considerable depth in relation to diameter which may be obtained by boring, drilling, reaming, tapping and trepanning (q.v.).

Degreasing Removing oil, grease, lubricant or greasy residues from a metal surface.

Detachable Boring Head A type of boring tool with a detachable head capable of being located wherever required along the bar length. It is highly flexible and will allow a plurality of cutters to be held. Some heads of this character are located on the end of the boring tool and will hold two or more cutters, which can be interchanged. Thus, the tool can be employed for a variety of bore diameters.

Diamond Edge Band Saws Band blades having a diamond edge, specially designed to cut exceptionally tough metals such as those containing comparatively high nickel or cobalt contents used for thermal resistance, and also for extremely hard steels. As these blades create a considerable amount of frictional heat, a cutting fluid is essential.

Diamond Grinding Tools Wheels coated with a diamond powder abrasive to remove material rapidly, but not for rough grinding. They give flat cutting faces free from rounded edges; produce smooth edges quickly and easily, without grinding cracks; and resharpen cutting tools at an exceptionally rapid rate. In some shops diamond grinding wheels are used for both roughing and finishing, but this is not usually recommended. Speeds of 3,500–4,500 surface ft./min. are advised.

Light cuts only are taken, and the wheel is often dressed (q.v.) to prevent overheating or surface cracking following on too high a wheel speed or too hard a wheel. Typical tool materials ground with these wheels include carbide and Stellite tools. Standard grit sizes are 80, 90, 100, 120, 150, 180, 220, 240, 320 and 400.

Diamond Tools Tools for cutting largely restricted to machining operations requiring the maximum degree of fine finish, though they may be used as chisels in tracer form-truing. They are normally supplied with either circular or many-sided cutting-edges, sometimes with up to five faces on a single cutting-edge. Parts produced by powder metallurgy are often turned or bored with these tools, which are also used to some extent for boring zinc, usually when the surface required has to be extremely smooth, or when a high precision machine is used to achieve extreme accuracy.

Diamond tools are costly and have to be lapped, but if correctly used will do a great amount of boring work without needing to be relapped. They are being increasingly used for the mass-production of components to precision tolerances with surfaces of highly-reflecting type. Diamond tools with a round nose cutting at low pressure and using light feed and shallow cuts give high production of light and non-ferrous alloy components to close limits.

The tools have to be held as rigidly as possible in the machine, and contact with the work is made as gently as possible. Once cutting has begun, the machine must not be stopped until the tool has been withdrawn from the work surface. If the cutting face of the tool shows any flaw or chip, it is taken out at once and relapped. No vibration or chatter at the tool nose should be allowed. A cutting fluid is not always essential, but a light oil emulsion is advantageous. Diamond tools are not employed for interrupted cuts.

For boring zinc alloy die castings the speed ranges from 800–1,500 surface ft./min., being the maximum possible that will not set up vibration or chatter. Feeds are fine (0·004–0·005 in./rev.) and depth of cut is a maximum of 0·008–0·025.

Diamond tools, broadly, are employed for special operations and will normally provide tools having a longer service life than those of tungsten carbide, especially for single-point turning.

Die Head or **Die Box** A cylinder with a shank to receive screwing dies for use in a screwing machine, the shank allowing the cylinder to be secured to the turret of a lathe. The cylinder body contains chasing tools for the forming of the threaded portion of the work. Various types of these tools are provided. Most die heads are self-opening so that the spindle need not be reversed for removal of the die. See **Threading**.

Die Head Supports Means of supporting the work when axial cutting is carried out in a die head. The heads used resemble those employed in **Die Threading** (see **Threading**).

Die Honing Guide A guide used in honing the bores of circular slitting saws of hardened steel, using a hand machine.

Die Threading See **Threading Dies**.

Dielectric Fluid A fluid made up of two parts of dielectric oil and one part of silicone fluid, used in clearing away the chips from the space between electrode and work in **electrical discharge machining** (q.v.). The fluid is passed down the hollow pin part of the electrode and constitutes the medium for electrical discharge.

Diemaker's Reamer A reaming tool of special form possessing a high helix angle and left-hand helical flutes, combined with a taper of about 0·013 in./in., employed for producing holes for dowel pins in parts of dies. See **Reamers**.

Dies Dies are of many kinds but, for the machinist, the principal types are steel blocks to which is given an internal thread whose cutting-edges produce screw threads with great speed and, if correctly manufactured and employed, with considerable accuracy. Dies have the advantage that they give a complete thread in a single cut on either cylindrical or tapering surfaces, but they do not form threads externally with the facility of a thread-rolling operation (q.v.). However, they form threads more quickly than the single-point cutting-tools sometimes used to produce screw threads, especially in hard metals, such as those above Rockwell C36. See **Threading Dies**, etc.

Diesinkers Machines that produce an impression in a die used for drop forging or similar shaping operations, and largely confined to one impression or a small number of impressions. The machine is vertical, small, and possesses a sliding table supported by a fixed and rigid bed. The vertical spindle is located on a pillar. Control of depth is automatically obtained for roughing-out, and for finish-machining a special mechanism is often added (see **Profile Milling**). Some machines have a servo-mechanism for giving particularly precise and largely remote control. These give various intricate cuts and forms, as for parts used in space applications.

Differential Indexing See **Indexing**.

Disc Cutter A type of cutter for machining gears as a generating operation. The cutter has teeth along its periphery and is located on a spindle capable of giving axial strokes while revolving, the work being mounted on another spindle in synchronism with that of the disc cutter, and revolving as the work is cut. At the same time the cutter is slowly fed into the work.

Disc Dampers An alternative to the **Damper Plug** (q.v.) for preventing or minimizing vibration in boring by inserting into the hollow of a boring tool a number of inertia discs whose diameters are not quite identical. If the boring tool vibrates, the discs slide one against the other with the result that they impinge upon the tool at random times, so lessening the vibration until it is too minute to produce chatter.

The discs are secured to each other by a bar passing through the shank and screwing into a threaded hole in the head of the cutter. This arrangement not only gives the assembly considerable strength, but also allows the heads to be replaced to suit different work and keeps the bar compressed.

DISC GRINDING

Disc Grinding A method of grinding to give a flat surface a rough or semi-precision finish. It removes material quickly and economically with somewhat less than maximum accuracy. Characteristic examples are grinding tools flat, squaring off the end of die blocks and grinding massive and heavy parts. The disc used is of abrasive material and is held in place by bolts or screws so that it is readily removed from the machine and replaced. It is strengthened by a steel plate at the spindle end, to withstand normal pressures even when grinding on the side rather than the edge. Surfaces of area up to 300 sq. in. can be ground in this way, as well as exceptionally small surfaces not above $\frac{1}{2}$ in. wide, for which a typical output of 2,000 pieces/hr./machine can be obtained.

The method will grind a larger area for an identical machine than a cylindrical wheel held in a chuck, as the disc diameter equals the chuck diameter and all the abrasive is used with no waste. The lighter wheels enable a better balance to be had and also greater rigidity, while overload on spindle and bearings is reduced. A smoother surface is also given to the work, facilitating truer location in the following operations. The method is extremely efficient and also serves to true-up plane surfaces.

The work is held against the side of the wheel and can be oscillated across it to prevent scoring of the wheel surface and minimizing wheel wear. Some work has to be clamped into place, but this necessitates a heavier cut to ensure truth of surface. Grinding speed lies between 5,500–7,500 surface ft./min. Higher speeds are possible, but may cause wheel fracture as a softer wheel is necessary, so that the lower speeds are preferable. These lie between 4,500 and 5,000 surface ft./min. and give greater accuracy and economy, though about 10 per cent more power is required. There is, however, no chatter, which reduces maintenance charges. Moreover, grit and grade of wheels can be more easily chosen, as the number of wheels completely safe to use is enlarged, and for this reason a wider range of materials can be ground.

The machines used are horizontal double-spindle, vertical-spindle and horizontal single-spindle. Machines can be obtained with automatic work oscillation, in which the reciprocating table concurrently travels a set distance towards the wheel, and when the stop shows that enough material has been removed, the feed and table motion is halted. In some of the powerful modern machines the work is located on a circular rotating work-table to give a high rate of production with great accuracy. The work is fed into the gap between the wheels on double spindle machines either manually, by rotary action or by reciprocation.

See Table IV for type, functions and speeds of suitable wheels.

TABLE IV
Disc Wheel Types and Speeds

Wheel Type	Speed (surface ft./min.)	Purpose
Solid resin and shellac	up to 9,000	All purposes
Segmental vitrified	up to 6,500	Heavy, dry grinding
Solid vitrified	up to 6,500	Wet grinding and light work
Solid silicate reinforced	up to 6,500	Dry or wet grinding
Plate-mounted	6,500–9,000 according to bond	All types of disc grinding

Dish Wheel Grinding A method of grinding metals in which a pair of dish-shaped wheels are employed. These are either parallel in a vertical position represented by 0 deg., or have an inclination of 15–20 deg. They are mostly used for grinding gears. The 0 deg. position is usually preferred to the 15–20 deg. because it gives a greater output with no loss of quality, and makes it possible to obtain more stringent lengthwise modifications and flank profiles. On the other hand some gears are less suitable for this position because there is no uniform change from flank to root of the tooth, which lack is not desirable with these gears.

Dividing Head See **Indexing.**

Dog, Lathe A type of clamp capable of being rotated by a lathe faceplate and used for holding work.

Dolomite Abrasive In abrasive-jet machining (q.v.) calcium magnesium carbonate or dolomite in a fineness of about 200 mesh, used for light cleaning and etching.

Double-cutting Tools Special tools used in planing (q.v.), and cutting on both the forward and backward strokes of the machine. They are held in toolholders of special design located on the planer head and themselves carried by an oscillating spindle which moves clockwise 13 deg. to set the tools in the proper position for cutting. At the end of the stroke, counterclockwise motion to the same degree places a second tool in position which cuts on the reverse stroke of the carriage or table. A standard finishing-tool is placed in the clapper box of the head to rough-cut and semi-finish a casting at one and the same time.

DOUBLE END SPOTFACERS

Double End Spotfacers See **Spotfacing**. These machines allow spotfacing and back spotfacing to be carried out on a pair of interior surfaces without taking out the cutter.

Double Housing Planers Planing machines (q.v.) embodying a pair of vertical columns to carry the cross rail. They have greater rigidity than open-sided planing machines, but the width of the work they can accept is limited.

Double-margin Drills A short and stubby design of twist drills with short flutes designed for cutting hardened 18–8 stainless and other difficult steels. The drills have exceptional rigidity, and for the best results are given a primary relief of 5–7 deg. flat, about 0·032 in. wide on the lip.

Double-roll Threading Attachments Threading attachments whose transverse pressure on the work is the least possible. They are used to the greatest effect in rolling threads at the collet ends of parts being machined in a lathe or automatic bar machine, and enable small diameters of more than usual length to be rolled at a longer distance from the collet than is possible with single-roll threading. See **Thread Rolling**.

Double-wheel Surface Grinders Machines for grinding components with opposing flat surfaces, but with the external edges of the opposing faces having the same form so that they give each other support and are uniformly ground as the work is admitted to and departs from the wheel faces. Small and medium-sized work is carried by rotary discs between the wheels, and larger work (above 4 in. wide), having about 20 sq. in. of area per face, is carried by shuttle carriers, which are less suitable for the smaller work. The wheels have high efficiency, but take longer to set up, so that a good quantity of work is required to justify their use and to justify the provision of the complex work-holders. Fine limits can be obtained and work to close limits of flatness achieved since there are no magnetic chucks with their typical holding and release forces involved.

Dovetail Tools Flat form tools giving greater rigidity because they are held in a special holder or adaptor. There are four types of these tools, all working in a radial direction. They are sometimes termed "prismatic" tools, but the term "tangential" is also applied on occasion, though erroneously. Some of these tools have the top cutting-face parallel to the machine-table, but others have the top cutting-face at an angle, so giving a top rake angle (q.v.). They have many advantages and a few disadvantages, all concisely summarized by Mr. W. F. Walker in his book *Form Tools* (Hutchinson, 1947). Their design to yield any required number of diameters or forms is not difficult, and many of them are tipped with either a super-high-speed tool steel or a cemented carbide. They are used to bevel the edges of inner and outer dovetails. Composite dovetail tools are not common.

Down Milling An alternative name for **Climb Milling** (q.v.).

Dressing Grinding Wheels Modification of the cutting-action of a grinding wheel by a sharpening process, not to be confused with **truing** (q.v.). The dressing tool is held in a tool-post or on a work-rest, or given rigid support. The work-rest is held as close as possible to the wheel, and the tool bears firmly against work-edge and work-rest. Space must, however, be allowed for the teeth, but if the space is too large the tool may be pulled into it, with disastrous results. The tool is traversed uniformly over the wheel face and not allowed to wander. Sparks indicate that both wheel and *tool* are being abraded, and when this occurs, the use of greater pressure forces the dressing tool teeth harder into the wheel surface, the heel of the tool bearing firmly against the work-rest edge.

Automatic traverse is advisable. The dressing is done by diamond or cylindrical grinding wheels, but small silicon carbide sticks (stones) lightly applied by hand are best for keeping the contour of fillets. For internal grinding wheels diamonds are also used, the silicon carbide stick smoothing off the rough edges. Surface-grinding wheels use diamond abrasive tools, which are the most satisfactory. Snagging (q.v.) employs a dresser specially designed for the purpose. Hand-dressing employs abrasive sticks mounted in wooden holders, or abrasive bricks.

The roughing-cut should not be greater than 0·001 in. Finish-grinding requires a slow feed and a light cut to give a smooth surface, finishing with a few passes across the wheel face without feed. The wheel revolves at the working speed, except in the case of thread-grinding wheels, which revolve more slowly.

Wheel-dressing tools made with small commercial diamonds set in a matrix of tungsten carbide are used in the United States. Solid copper diamond dressing tools are also manufactured, and their thermal conductivity is said to disperse the heat rapidly.

Drill Press A loose term commonly used for any type of machine suitable for drilling metal, probably because pressure is applied to force the tool through the metal as it rotates.

Drilling A method of boring holes in metal or other materials with a rotating tool cutting on the end and provided with suitable cutting edges or lips to which the proper angles are given, as well as a number of longitudinal flutes designed to allow chips as well as a coolant to pass along them. It is commonly regarded as the most effective

DRILLING

commercial means of forming holes accurate to dimensions in solid material.

The drill may be used in a lathe, drill-press, radial drilling machine, turret lathe, gang-drilling machine, multiple-spindle machine, turret drill, horizontal drill and automatic drill, while some drills can be used economically in a portable machine. These machines are all dealt with under their own names.

The two principal types of drive are straight shanks (q.v.) for small drills, i.e. those of small diameter, and Morse taper shanks (q.v.) for the larger drills, while there are also a few square taper-shank drills used in ratchet braces. Straight shanks are popular up to about $\frac{3}{8}$ in. dia. because they are inexpensive. The taper shanks are clumsy for small sizes, and seldom used below $\frac{3}{16}$ in. All drills above $\frac{1}{2}$ in. dia. are better made with Morse taper shanks, which hold the tool more rigidly, allow of a quicker tool change, give greater concentricity and closer spacing when drills are used in a "gang", and enable the tool to be driven without slipping in the socket. They are easier to remove from the socket and rarely get out of alignment.

Shallow holes may be drilled with straight-shank drills because money is thereby saved, but the straight-shank when used on the deeper holes may be forced out of the socket on the spindle return stroke, which means that the work must halt.

The drills themselves are usually made of a high quality carbon steel for the straight-shank small diameter drills, and high speed steel for the taper-shank drills. Some drills in sizes from $\frac{3}{8}$–$2\frac{3}{8}$ in. dia. and with taper shanks are often made with shanks of oil-toughened special alloy steel welded to a high speed steel cutting portion. This is because the special alloy steel is hardened and toughened in oil to offer maximum resistance to the torque (q.v.) or twisting moment (action) encountered in drilling plate, etc. These do not readily burr over when severe and repeated drifting takes place. The welding is by electrical butt-welding process.

Twist drills are normally ground with lip clearance, equal lip-angles and lengths and point-thinning. See Fig. 13. The lip corners are sometimes rounded on difficult jobs to prevent heavy break-through stresses from causing the corners to chip. The rounding helps to distribute both the cut and the frictional heat of cutting, so helping to prevent a slackening of speed, but this treatment is not advised for the drilling of austenitic manganese steel (12–14 per cent manganese). Typical speeds and feeds for drills are shown in Table V (page 69).

Drills tipped with tungsten carbide are increasingly used for

DRILLING

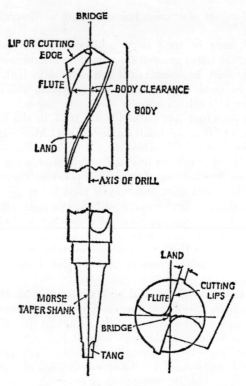

Fig. 13   Twist drill nomenclature

drilling abrasive materials of low tensile strength, e.g. cast iron and castings of high silicon aluminium alloys or heat-resisting alloys. They are also used for drilling hard steel (over Rockwell C48), but this use is not widespread. Porcelain, glass, bakelite and other materials have been successfully drilled with these tipped drills. A few carbide drills are made solid for work on parts required to show holes of great accuracy and where rigidity of the tool is essential.

Some drills are made with specially short and stubby cutting portions, i.e. with shorter cutting flutes. On difficult work such as drilling austenitic stainless steel plate these drills enormously increase the margin of safety, eliminate drill fractures and blind holes, and increase the average number of holes to over 100 per grind, reducing the cost of the drills and increasing production with an economy that outweighs the cost of cutting down the drill shanks.

Drills used in a horizontal machine may prove difficult to lubri-

"DRY"

cate when cutting deep holes, and therefore some users employ an oil-tube twist drill with a copper tube inserted centrally in the body behind the lands or by drilling holes longitudinally in the solid body, which is better. The drilling of the holes is done before the twist is given to the drill. The coolant is supplied under pressure to the drill spindle, flows through the shank to the lips, then back along the flutes, taking the heat and chips away with it as it goes. This type of drill is also suitable for drilling deep holes vertically. Drills with a taper shank render the holes more concentric, enable speed and feed to be increased and give a more rigid drill.

TABLE V
Drilling Speeds and Feeds (typical)

Material	Speeds (surface ft./min.)	Feeds (in./rev.)
Steel castings	36–80	0·008–0·08
Steels (low carbon)	32–40	0·008–0·06
Tool Steels	20–33	0·008–0·02

Some drills have a slow spiral and a relatively thin web, and are particularly suitable for drilling plastic materials, brass and soft materials because they enable the chips to be cleared with ease, and break them up into small particles. They have greater rigidity than fast-spiral drills and withstand higher torque stresses, while they do not so often drive right through after the drill point emerges from the material. Drilling shallow holes in steel and alloys of magnesium or aluminium is another use for these drills.

Fast-spiral drills on the other hand have wide flutes and narrow lands, and are for work on deep holes in non-ferrous alloys, and also for drilling carbon, alloy and stainless steels. They facilitate chip removal and increase the drill-bearing surface.

Left-hand drills, in which the direction of the spiral or helix is reversed, are used for multiple-working machines in which the spindle revolves in the opposite direction to the usual, as when tapping succeeds drilling or when other operations are combined with drilling.

Some special drills are dealt with under separate headings, e.g. **Gun Drills, Core Drills, Jobbers' Drills, Screw Machine Drills, Step Drills,** and **Spade Drills.**

"Dry" In the machining of certain metals that rapidly work-harden on the surface as soon as cut or abraded, it is necessary to generate frictional heat in order to soften the work-hardened skin. This enables the tool to get underneath the skin and remove it without difficulty. For this reason it is advantageous in some instances to machine without a lubricant.

Dry Drilling Drilling without the use of any coolant or lubricant, as when drilling work-hardened austenitic manganese steel (12-14 per cent manganese).

Dry Grinding Grinding without coolant or lubricant, sometimes adopted for cast iron as a means of giving a better finish and lengthening grinding wheel service-life. Other materials sometimes ground dry are heat-resisting alloys, magnesium alloys and tool steels.

Dry Reaming Reaming without the aid of a cutting fluid, usually when grey cast iron is being cut. An air-jet is used to cool the work and clear the chips.

Dry Turning Turning when no coolant or cutting fluid is used, mostly in machining cast iron and austenitic manganese steel.

Dual-ram Broaching Machines A type of broaching machine able to perform either a single or a dual broaching operation.

Duplex-face Thread-rolling Dies Dies for forming straight threads threaded on both front and back, giving two rolling surfaces. If the screw length is below half the width of the die, top and bottom are reversible, so providing four rolling edges and giving maximum economy. (See **Thread Rolling**.)

Duplex Milling Machines See **Milling Machines**.

Duplicating Lathes A generic term covering copying, tracer, profiling, numerical control and continuous-path turning lathes. (See **Lathes**.)

E

Edge Build-up on Cutting Tools When a tool is cutting, the particles or pieces of metal it shears off are extremely hot owing to the frictional heat generated. In consequence they are liable to weld themselves to the edge of the cutting tool, causing the work to have a rough surface. The degree and rapidity of build-up depend on the tool material used, being faster for high speed steel, and also on the cut depth and tool nose radius. Carbide tools usually cut at so high a speed that the chips do not readily accumulate or weld. See **Chipbreakers**.

Edge Locator. A form of tracer for fixing the location of the axis of a machine spindle in relation to the datum edge of a workpiece.

Edge Profiler An automatic fixed spindle machine having an integral power-feed unit for the quick and accurate jig machining of the external faces of components.

Electrical Discharge Grinding See also **Electrical Discharge Machining**. The grinding of a metal by the action of electrical sparks flowing in a stream between a grinding wheel, negatively charged with electricity, and the work, positively charged. The work is held in a **dielectric fluid** (q.v.). The stream of sparks when the work is carried to the wheel, which is of graphite, causes a minute quantity of metal from the surface of the work to liquefy or vaporize, leaving behind a tiny crater. The sparks are produced by d.c. in high frequency pulses. Less current is employed than for electrical discharge machining owing to the much smaller cutting area. This form of grinding gives exceptional precision and an excellent ground surface. The wheel revolves at from 100–600 surface ft./min. The spark gap is usually 0·0005–0·0030 in., and the work table is servocontrolled (q.v.).

The method is employed in producing form tools. Tolerances obtainable may be as low as ±0·002 to 0·00005 in. The operation is slow, the rate of metal removal being from 0·01–0·15 cu. in./hr. Higher rates mean rougher surfaces. Typical applications are to carbide form-tools, dies, rolls, etc., hard metal gear-racks, thin closely-spaced slots in hard materials, low ductility or delicate components, and producing intricate forms at a high rate of output. It is not suitable for cast iron.

Electrical Discharge Machining (Electro-spark Cutting) A system of producing holes, slots or other recesses in material of electrically-conductive type by the action of a stream of electrical sparks at high frequency in liquefying or vaporizing metal. The work is usually given a positive electrical charge by regulated d.c. pulsing, and the tool is negatively charged. Thus an anode and a cathode are separated by a space measuring about 0·0005–0·020 in. A dielectric fluid is employed to immerse both work and tool. The net result is the formation of a minute crater in the work surface after the sparks have passed. See also **Electrical Discharge Grinding**.

The technique is primarily applied to machining dies and moulds in advance of or following upon hardening, cemented carbides, tungsten, honeycomb structures and other delicate materials. It will also produce small holes of considerable depth or irregularly-formed holes down to 0·001 in. (which it would be almost impracticable to produce by alternative processes), as well as small, complicated or exceptionally accurate components. The operation is heavy on tools which are high in first cost. Hence, no special advantage is derived in normal machining with these machines, but given suitable graphite electrodes, some economy is achieved when forming contoured recesses of considerable size in hard steel dies and parts. In general the process surpasses ordinary machining for work of great hardness, high tensile strength, low machinability, intricate or difficult form or delicate construction, having to be free from rough edges and needed for the close mating of a punch and die. Holes with a 20:1 depth-to-diameter ratio may be formed, as well as narrow or awkwardly formed slots, holes or slots having acute angles and required in considerable numbers.

Sometimes the electrodes are positive in polarity, as with the steel, aluminium alloy or copper-tungsten electrodes used in machining steel, and the graphite electrodes used in roughing. The machines employed are of ram type, the work-head being given its motion by a hydraulic cylinder, but a machine with an auxiliary spindle operated by a hydraulic motor driving bevel gears and a lead-screw, or by a hydraulic cylinder, is used when the components are not of large dimensions, being more economical per piece machined. The tool is moved forward by a servo-mechanism electrically actuated, which maintains the distance between anode and cathode without variation.

Dielectric fluids such as a hydrocarbon petroleum oil with adequate viscosity, paraffin, silicone oils, de-ionized water and water solutions of ethylene glycols, are passed under pressure between anode and cathode through either a hollow tool or outside jets or orifices in the component. This fluid must have low viscosity, a high ignition temperature and be inexpensive. It acts as a conductor of the spark flow and as a chip remover, and is usually filtered to eliminate unwanted matter and machining particles.

In the most modern machines the power is supplied by a vacuum tube pulse circuit, the charging being but a minor part of the cycle. The circuits are somewhat limited, but the limitations are overcome by modifying them in certain directions. Control for any specific machine, electrode and dielectric fluid is achieved by modifying the current, the period of discharge, the pulse frequency, and the distance between anode and cathode or *spark-gap*.

This form of machining has been applied to aluminium alloys, heat-resisting alloys, nickel alloys, stainless steels, alloy steels, titanium alloys, tool steels and tungsten.

Electrical Discharge Trepanning See **Electrical Discharge Machining** and **Trepanning**.

Electrochemical Discharge Grinding A form of grinding which employs in combination slightly varied forms of **Electrochemical Grinding** (q.v.) and **Electrical Discharge Grinding** (q.v.). It is employed primarily for grinding cemented carbide tools, the harder tool steels, alloys having a nickel base and delicate components liable to fracture or greatly affected by high temperatures. The technique involves oxides being produced on the work, which constitute the anode and are positively charged by an a.c. or pulsing d.c. No abrasive wheel is needed. The spark discharges are discontinuous, but effectively remove metal from the surface of the parts. No dielectric fluid is used, and the current is low voltage h.f. An electrolyte with high conductivity is used instead, usually an inorganic salt of anhydrous type (sp. gr. 1·11–1·15) about $1\frac{1}{2}$–2 lb. gall.

Anodic oxidation in the electrolyte is promoted when unbroken and adherent films of oxide are electrochemically produced on the surface of the work and eliminated without interruption at a fast enough rate. In other words, the metal is at first changed into an unbroken, low-conductivity film of oxide about 10–50 micro-in. thick. When the work has a positive charge, given by a.c., the film thickens and is eliminated at the point of discharge as soon as the breakdown voltage of the film is surpassed during the ensuing negative pulse, whereupon the cycle is repeated. If pulsing d.c. is used, the film thickens during the low voltage part of the pulse and is eliminated at the discharge points. Given a few seconds, the discharge points are scattered haphazardly over the area to be machined.

The wheel used does no grinding and is usually of a solid bonded graphite. The work is not cratered, though the wheel is, at the points of spark formation, the extent governing the rate of wheel wear. The wheel runs at 4,000–6,000 surface ft./min.

As compared to electrical discharge grinding, the wheels cost considerably less, are easily given complex forms and contours, and can be dressed in position, while their period of service-life is longer. The number stocked is much lower, as one type only is required for a number of different machining operations. The operation uses from 10–15 times as much current when used at the same rate, but eliminates far more material at a more rapid rate and with a better surface, though accuracy to dimensions is less. The work is applied to the wheel at a much reduced pressure, so that more delicate operations can be carried out on various structures difficult to grind by alternative means.

Electrochemical Grinding This grinding operation combines ordinary abrasive wheel grinding with electrochemical action. The abrasive wheel is responsible for the elimination of a mere 10 per cent of the material, the rest being taken off by the same method as in electrochemical machining (q.v.). The parts to be machine-ground are not passed at high pressure between anode and cathode, but are bathed in an electrolyte. The wheel for grinding propels the electrolyte in spray form. The current travels from the power source (d.c.) through the positive anodic work and the electrotype to the wheel, which constitutes a negative cathode, and so returns to the source. The spray is collected up and recirculated by way of a supply container after filtration.

The metal passes into solution in the form of metallic ions, globules of hydrogen gas being generated at the wheel. Wheel and spindle or work are insulated. The projecting portions of the non-conductive abrasive material ensure that the wheel body, which is conductive, does not cause a short circuit against the work, and also control the gap for cutting. This must be completely filled with electrotype while the operation continues or the metal will not be removed at the proper rate and the wheel will be worn too heavily.

The electrotype is usually a sodium or potassium salt (1–2 lb./gall.) giving 0·05–0·15 mho/cm. Less power is often needed for electrochemical grinding than for electrochemical machining, the surface to be ground being smaller. Current lies between 50–3,000 amp. at 4–10 volts through a gap of about 0·001 in. The wheel revolves at 4,000–6,000 surface ft./min. at ordinary room temperature. It is specially made up for this class of work, and is of aluminium oxide type with a bonding agent. Diamond can also be used as the abrasive medium. The wheelmakers recommend the proper grit and grain sizes for these operations. There are metal-bond and resin-bond wheels. Pressure against the work ranges from 50–200 lb./sq. in.

Accuracy up to 0·0005 in. can be obtained given suitable conditions, but not on sharp corners. Sometimes this can be even greater if a last mechanical finishing-grind is given with the current shut off. Rate of cutting is governed by current density, which depends on the type of material being ground, and which may range from 500–4,000 amp./sq. in.

The process is primarily used for tungsten and other cemented carbides, heat-resistant alloys, nickel alloys, stainless steels and tools.
Electrochemical Honing See **Honing.** In this process a honing effect is produced by combining anodic dissolution of the metal with mechanical abrasion. As compared to mechanical honing, it eliminates material at a faster rate, minimizes the formation of burrs and gives a longer service life to the bonded abrasive, while the sur-

faces finally produced are almost entirely without injury or stresses resulting from heat generation. Electrochemical honing will do whatever mechanical honing will, and is most successfully used for hard metals. It corresponds largely to electrochemical machining, but allows a wide choice of electrolyte, though sodium chloride solution is employed in most instances. The greater part of the material is eliminated by electrolytic action, the work constituting the anode and the metal tool the cathode. There is a gap of 0·003–0·005 in. separating them at the outset, but this enlarges up to 0·020 in. and above as material is taken off.

The honing tool is of stainless steel, precision-bored, revolving and reciprocating on a rigid spindle, the electrolyte being passed through the tool and discharged through orifices in the tool body into the gap. The honing is done by bonded abrasive stones (sticks) placed on the tool and emerging through slots in its body, being thrust outward by a conical piece in the tool so that they press without interruption against the work surface. The stones, numbering three or more, eliminate the material of the work by mechanical abrasion and by allowing it to offer new surfaces to the action of the electrolyte, so ensuring a precision bore with the least possible loss of metal. The stones and their bond have to withstand the chemical attack of the electrolyte.

Up to now only internal cylindrical honing is being done. Current density is usually 120–130 amp./sq. in., but may be as much as 300. Electrolyte pressure is about 150 lb./sq. in. When the work is almost completed the current is switched off and the honing stones continue to act for a few seconds longer. The maximum honing length of bore is 12 in., and the diameters may range from about $\frac{3}{8}$–6 in. The electrolyte is collected, filtered and recirculated. (See also **Electrochemical Machining.**)

Electrochemical Machining Eliminating undesired metal from a work surface by dissolving it anodically in an electrolyte, using the work as anode and the tool as cathode. The electrolyte is delivered under pressure into the space separating work and tool, d.c. being passed through at low voltage.

The operation is designed for machining parts that cannot be easily machined by machine tools. It is applied mostly to hardened steels, heat-resistant alloys, face-milling, deburring, etching, marking, and the production of irregularly formed holes of small diameter and considerable length in relation to that diameter. It achieves its maximum economy when producing great quantities of identical components as it is expensive in first cost, setting up and tool cost. Principally, therefore, its use is confined to parts for space vehicles,

ELECTROLYTIC GRINDING

jet engines, rocketry and automobile manufacture, though it is also used to some extent in general production.

The electrolyte used is a water-soluble inorganic salt combination, such as sodium and potassium chloride, but sulphuric acid and sodium hydroxide solutions have also been used for specific jobs. The current density is 100–2,000 amp./sq. in., current being 50–20,000 amp. at 4–30 volts d.c. The gap ranges from 0·001–0·030 in. Electrolyte flow is 20–200 ft./sec. with pressure 10–400 lb./sq. in. Electrolyte temperature is 25–65 deg. C. (75–150 deg. F.). The fluid is filtered and recirculated after use.

Tool motion is regulated by a servo-mechanism. Tool feed rate is constant in relation to metal removal and must therefore be uniform and easily regulated. Smaller holes call for faster feeds, and given unchanged current, feed changes modify cut dimensions. Most machining by this process uses a constant voltage with a fixed rate of feed. Penetration rates range from 0·25–6·0 in./min. according to the type of operation. Current density chiefly determines the allowable feed of tool, and governs also the work finish, the higher the density the greater being the polish, other factors being equal.

The tools must be rigid, machinable, conductive of heat and electricity, and chemically capable of withstanding attack by the electrolyte. Copper and brass are the most suitable, except when rigidity is the primary need, when stainless steel or titanium may be used. They are given a high degree of polish to prevent marks on the work surface, and on occasion chemical polishing is also given after mechanical polishing. Insulation of the tools to regulate the current path is essential. A wide range of insulating materials can be used, such as nylon, acetal and fibreglass-reinforced epoxy, but other materials can also be used with success in suitable conditions. (See also **Embossing by Electrochemical Machining**.)

Electrolytic Grinding See **Electrochemical Grinding**.

Electron Beam Machining A machining process in which high velocity electrons are focussed or "beamed" on a material surrounded by a vacuum. Their kinetic energy is then transformed into heat, and vaporization of a minute portion of the material takes place. The work is done in a vacuum chamber to obviate dispersal of the electrons by collision with molecules of gas.

The velocity of the electrons should be about 50 per cent of the speed of light, so that they have to be speeded up to attain this. They immediately reduce any material to a vapour, and are projected at 10^{10} watts/sq. in. The area of the work on to which they are focussed is 0·0005–0·001 sq. in., which dimension is attained with absolute precision.

The operation is for work of a thickness between 0·010 and 0·250 in., and will produce holes as small as 0·0005 in. dia. irrespective of the work material, while slots no wider than 0·001 in. can be machined with no more than 0·005 in. between them. Another function is to dislodge taps of small dimension from holes in which they have become lodged through fracture. The rate of penetration is around 0·010 in./sec., but faster rates are possible.

The machine is made up of a vacuum chamber having a vacuum of the order of 10^{-5} mm. Hg in which is a cathode negative electron-delivering tungsten filament at a temperature of about 2,500 deg. C. (4,530 deg. F.). A grid cup forming the lower end of the cathode develops a magnetic field, under whose influence the mass of electrons is formed into a tapering stream beamed onto the positive anode, with which it makes no contact. Instead, it passes through an aperture in the anode, and the difference in potential between cathode and anode, amounting to 50–150 volts, provides the necessary acceleration.

On leaving the anode aperture the electrons attain their highest velocity, which, since there is no interruption or retardation by gas molecules in the vacuum chamber, they sustain when they impinge upon the work. The beam, which is therefore electromagnetically projected, locally heats the area of the work it encounters and causes vaporization. The anode is in effect the gun that aims them, and is about 4 in. from the work. The current of the beam ranges from 100–1,000 micro-amp., with continuous power of 100–1,150 watts. The pulse duration is between 4 and 64,000 micro-secs. at 0·1–16,000 c.p.s. frequency.

The focussing of the beam is achieved by a magnetic lens, giving the beam a deflection to any point within a work area of $\frac{1}{4}$ in. sq. Magnetic coils located under the lens enable the electrons to machine in various configurations from a minute round to a rectangle or square. The beam is regulated in conformity with the material to be machined, the form and dimensions of the cut. The penetration of heat below the work surface is from 0·001–0·010.

The operation drills exceptionally small holes with extreme accuracy in a short space of time irrespective of the work material, and when no other process can achieve the same result. Slots, etc., can also be produced. There is no contact between tool and work, and automatic machining by the method is perfectly feasible. On the other hand the apparatus is costly, and no cut can exceed $\frac{1}{4}$ in. The dimensions of the work are governed by those of the vacuum chamber and the amount of material that can be machined off is limited. The time occupied in removing a given amount of metal and producing the desired degree of vacuum in the chamber is protracted. Skilled operators are required and the holes or slots produced are by no means invariably uniform, some having craters at their mouths and in any event tending to lose diameter as they proceed. The larger the hole or slot, the less accurate.

In general an accuracy of ± 0.0002 in. can be had with cuts from 0·020–0·040 in. in depth. The beam density at the tip of the cathode does not exceed 14 amp./sq. in. and the temperature at the tip, using a tungsten cathode, is 2,525 deg. C. (4,580 deg. F.), which is a maximum, as above this temperature the service life of the cathode rapidly declines.

Beam motion is possible within about 0·040 in. Usually a microscope enables the work to be examined during operation. Heat penetration liable to impair the material is much less in electronic welding. Pulses of brief duration prevent heat damage as far as possible, but necessitate a higher current. Typical work to which the operation is applied includes parts of aluminium, nickel, titanium, iron, molybdenum and tungsten.

Electro-spark Cutting See **Electrical Discharge Machining**.

Embossing, Electrochemical See **Electrochemical Machining**. In this operation both tool and work are held motionless in a solution of sodium chlorate constituting the electrolyte, the work being thus given greater precision in form and measurements and a superior finish of surface. No portion of the work has to be sealed off nor has the tool to be treated in this way. The operation is specially advantageous on hard steel. The electrolyte is strongly oxidizing, so that it must not make contact with materials which ignite readily, but it does not corrode neighbouring regions of the work.

Emery A medium soft natural abrasive used in grinding, lapping, etc., made up of a mixture of alumina and iron oxide. It is in effect a less pure form of corundum and is tough, durable, but not now favoured for abrasive wheels.

Emulsions See **Cutting Fluids**.

End-cutting Reamers Tools for finishing blind holes where it is important not to have a radius at the bottom, or to have one that is as small as possible. The tools are also used to correct deviations of up to several thousandths of an inch from the parallel in a hole drilled right through. They may have either straight or helical flutes, but their ends do not carry a bevelled portion leading the tool into the work. The surface they produce being somewhat rough, they are often followed up by a finishing reamer to give the desired smoothness of surface. See also **End Mills** and **Reamers**.

End Milling A machining operation of **Milling** (q.v.) type carried out with a tool having cutting-edges on the end of the face as well as on the circumference, producing cuts of considerable difference in contour by employing both side and end, either in succession or at the same time. It is less efficient than alternative forms of milling because the cutter-end lacks support and there is an unduly high proportion of length to diameter, so that it is not possible to make cuts of considerable depth. Nevertheless, the operation is often used for facing, recessing, slotting and box milling. See **End Mills.** Solid carbide tools are sometimes used to reduce the degree of tool deflection, and other expedients for the same purpose include electronically regulated feed, very light cuts, and adaptors of collet type to give the end mill maximum security. Minimizing run-out by means of an indicator also lengthens tool service life, enabling a higher degree of finish to be obtained.

End Mills See **End Milling.** Tools having their cutting-edges or teeth at the end and also on the circumference, and made as short as possible. They may be either right- or left-hand, and in most instances have helical teeth or flutes. Their function is to mill those surfaces that cannot be readily dealt with by cutters carried on an arbour. Another type of end mill has two fluted cutting-edges and mills out a groove, taking shallow cuts. Owing to its form, it will penetrate the work without the necessity of starting the cut with a drill. Standard tools of this type range from $\frac{1}{4}$ in. and above to $1\frac{1}{2}$ in. dia. The commoner multifluted end is used for a wider range of operations, and will take comparatively deep cuts.

The mills have tapered shanks, smaller in diameter than the receiving hole in the milling machine spindle, so that they have to be inserted in an adaptor. The mills are also made with straight shanks. (See also **Shell End Mills.**) Standard end mills are usually obtainable in diameters from $\frac{3}{16}$–4 in., but larger diameters can be had. Those below 2 in. dia. are normally made of solid tool steel, whereas the larger ones have inserted teeth of tool steel or tungsten carbide, usually with O-positive rake and positive axial rake.

Engine Lathes See **Lathes.**
Etching See **Chemical Machining.**
Expandable Shell Reamers See **Shell Reamers.**
Expanding Gauge. A type of gauge used in honing, made up of a split sleeve held in place by a ring and though not secured to the tool, reciprocating with it. The diameter of the gauge is less than that of the diameter to be gauged, and the tool penetrates the bore with each downward stroke, at the bottom of which a lever on the side of the sleeve engages a post and expands to establish contact with the hole surface. By careful previous setting of controlling de-

vices, a couple of electical contacts on the sleeve lever come together and arrest the cycle as soon as the gauge diameter equals that of the hole. The gauge is mainly used for regulating the sizes of holes above 0·75 in. dia. to within 0·010 in. by means of a calibrated dial, any setting of which enables a limit of 0·0003 in. to be maintained.

Expanding Pin Reamers Reaming tools in which the size is modified by an expanding pin with a screw thread, which causes the blades to move. They have straight flutes and are obtainable in sizes from $\frac{7}{16}$–$2\frac{1}{2}$ in. dia. No tapering flutes are available. The expansion allowable is $\frac{1}{32}$ in. for dias. of $\frac{7}{16}$–$1\frac{1}{4}$ in. and about $\frac{1}{4}$ in. for dias. $1\frac{1}{4}$–$2\frac{1}{2}$ in. They are often more profitable than solid reamers, as the blade may be reground many times before being scrapped, while the small adjustments give the tools longer service life between grinds. The larger adjustments enable the same tool to produce holes of somewhat different diameters, which reduces the number of reamers kept in stock. This is specially advantageous when only a small number of holes have to be reamed. The blade materials can be altered as required without change of tool body, and in regrinding design is more readily modified as regards angles, margins, clearance, etc., than with a solid reamer.

Expansion Cones Tools used for the honing of large bores in power stroke fixtures, when feed-out of the honing stones is required. (See **Honing**).

Expansion Taps Taps employed in finishing operations in tapping (q.v.) or for tapping metals having better than usual machining properties. They are somewhat expensive, and include an axial hole drilled from front to back of the tap thread portion and beyond, and a radial slot or series of slots or saw-cuts running from tap surface to axial hole between each couple of flutes. The slots give flexibility to the threaded section.

External Broaching Machines See **Broaching Machines**.
External Honing See **Honing**.

F

Face Angles See **Tool Angles**.
Face Chuck See **Face-plate**.

Face-grinding Machine An important design of grinding machine possessing a horizontal spindle, the work being carried by a vertical work-table and ground with abrasive wheels either segmented or cylindrical. The machines are of considerable dimensions and either work or wheel may travel.

Face-milling See **Milling.** A form of milling designed for cutting flat surfaces with the aid of a milling cutter whose cutting teeth run round its circumference and whose spindle drives the tool at an axis perpendicular to the work. The cuts it makes are radially deep and axially narrow, so that the machine needs less power to remove the same amount of material than a cutter whose teeth cut circumferentially, and which gives shallow radial and wide axial cuts. In addition the tool is more rigid, being securely fixed to the end of the spindle, while considerable areas can be tooled with only a minor projection of the spindle. The power in cutting is uniformly distributed, tools are taken out and replaced more rapidly, and the operation expense is lower, while the dimensions of the work are less restricted. Metals economically cut by this method include aluminium alloys, beryllium, cast iron, copper alloys, hardened steel, magnesium alloys, stainless steel, titanium alloys and tool steels.

Face-milling Cutters Cutters for milling whose teeth are on the face of the cutter or on the cylindrical surface. The teeth may run parallel to the cutter axis or at an angle to it, or the cutting edges may be helical on the cutter surface. Rectangular grooves can also be milled by a cutter with teeth on the circumference and both faces. Special cutters of this type are made for shaping grooves or depressions of particular forms, and are used for milling flat surfaces. In general, the cutter face finishes the surface, although the chamfered cutting edges take off the greater part of the metal. The spindle axis is perpendicular to the work surface. The dimensions of the cutters range from $2\frac{3}{4}$–20 in., the sizes above 3 in. dia. having inserted teeth. (See **Face-milling.**)

Faceplate (sometimes **Face Chuck**) A metallic slotted plate, for holding work that cannot be held in a chuck, introduced between a flange and the mandrel nose of a lathe. The faceplate is secured by screws, bolts or taper plugs to the flange and to the nose. The more usual application of the term is, however, to the normal lathe chuck slotted or bored to receive T-slots or bolts, or having movable individual dogs or jaws secured to the plate. Such a faceplate is often used in lathes when simultaneously cutting multiple threads, the finishing cut being made in most instances with a single tool.

Fast-spiral Drills (See also **Drilling.**) Twist drills whose flutes have greater width than the standard, and correspondingly narrow lands. The flutes are helical and the helix angle is high. They are used for drilling relatively deep holes in metals of non-ferrous type, as well as for carbon, alloy and stainless steels. The combination of high helix angle and wide flutes clears away the chips with great facility and provides a greater surface of land to each inch of length, so that the drill's bearing surface is greater. As these tools are usually fed in at a high rate, they are able to clear chips from holes of considerable depth much more effectively than the standard and other drills.

Feeds and Speeds Operators are always puzzled and disappointed when a tool manufacturer declines to indicate the "best" speeds and feeds for particular tools, except as a rough guide whose limits are wide. There are so many variables for every cutting operation that only the operator himself can assess the governing factors clearly enough to decide, usually by a judicious blend of recorded experience and trial, the exact feed and speed for a specific operation.

Feeds and speeds are affected by: (a) the type and kind of material to be machined; (b) the operator's skill and experience; (c) the rate of output required; (d) the age, type, power and condition of the machine tool; (e) the heat-treatment given to the cutting tool before and during use; (f) the type of cutting fluid used; (g) the design and size of the cutting tools; (h) the type of operation, the amount of material to be machined off, and the area to be cut; (i) the material of which the tools are made; (j) the number of times the tool-cutting-edges will need to be reground in relation to the output and the efficiency of re-grinding; (k) the surface finish required; (l) the limits of cost for the operation; (m) the form and intricacy of the work; (n) the ratio of cut depth to rate of feed.

Table VI may help in choosing suitable feeds and speeds as long as it is borne in mind that the figures given are *rough guides*, not precise recommendations, and are a basis for experiment, not a substitute for it.

Feeds of Tools See **Feeds and Speeds**.
Feeler Gauge A thickness gauge embodying a range of blades of varying thicknesses. See **Gauges**.
Felt Polishing Wheels. Polishing wheels made of circular pieces of woven wool-felt in different grades, from fine to coarse, and given a coating of abrasive material. They are claimed to give a more rigid wheel than one made of cloth, the circular pieces being thicker, flatter on the face and less liable to become radiused on the corners. The wheel face when coated and solid preserves its "give" and is all in one piece, not divided into separate discs.

Speeds and Feeds
Table VI

Operation	Cutting speed (feet per minute)				Steel Castings	Machinery Steels		Tool Steels
	Steel Castings	Machinery Steels		Tool Steels		22–38 tons tensile	38–63 tons tensile	
		22–38 tons tensile	38–63 tons tensile					
						Feeds (in. per rev. or per stroke)		
Lathe								
Turning	—	—	—	26–40	0·012–0·2	0·2–0·08	0·04–0·02	—
Roughing	20–50	100–80	60–40	—	—	—	—	—
Finishing	25–60	120–90	70–50	35–50	—	0·008–0·006	0·004–0·002	—
Threading	13–26	40–33	26–16	10–16	—	—	—	—
Recessing and Cutting off	13–40	66–59	46–40	20–33	0·008–0·04	0·004–0·04	—	—
Planer								
Planing	25–50	53–72	40–32	20–33	0·01–0·16	0·02–0·1	0·06–0·02	0·02–0·04
Slotting						Feeds (in. per rev.)		
Drilling Machine								
Twist Drill	36–80	82–65	50–40	20–33	0·008–0·08	0·008–0·06	0·008–0·044	0·004–0·02
Boring Bar	20–70	80–60	50–40	20–33	0·008–0·06	0·008–0·04	0·008–0·02	—
Counterboring	13–40	26–33	26–7	15–25	0·008–0·04	0·008–0·04	—	0·008–0·02
Milling Machine						Feeds (in. per minute)		
Long and Plain Milling	26–60	72–46	46–20	15–33	1–8	1–8	—	1–2½
Round Milling	26–60	80–60	45–30	15–33	¾–2½	½–2	—	½–1½
Tooth Milling	35–60	80–60	45–30	20–40	¾–2¼	⅝–2½	—	½–1½
Thread Milling	26–52	65–50	40–26	15–26	—	—	—	—

Ferric Oxide A soft grinding abrasive with a grit size of 1 micron, used in lapping soft metals to give a good degree of polish, or to give a final lap to parts whose surfaces have to reflect light to some degree. The amount of metal removed is extremely small and the reflectivity low.

Fillet Rolling See also **Roller Burnishing**. An operation employing a formed narrow roller for rolling fillets inexpensively under a degree of pressure.

Finishing Tools Cutting-tools used in lathes, etc., for making the final cut on the surfaces of steel or metal parts to bring them to the desired dimensions. They normally have wide, straight cutting-edges and eliminate the marks left by the tools used for rough turning. They include side tools, either left or right hand, square tools and spring tools.

Fishtail Cutter An uncomplicated type of milling cutter for work on a shaft or other part. When run at high speed with a light cut and feed it is used for producing a keyway, groove or seat.

Fixtures Accessories for setting up work to be machined, often rendering the operations more economical. Special clamps, for instance, speed up setting because they can be quickly tightened or released and are powerful enough to maintain work or jig in place, but must be properly used. Clamps or thrust blocks are often fixed to stop the tool or cutter from pushing the work along. Tool-setting is not fully controlled by fixtures, which are more commonly used in milling and turning operations. Both simple and intricate fixtures have been designed to suit particular operations either simultaneously or when indexed. (See **Jigs.**)

Flanged Chuck See **Faceplate**.

Flash Point The temperature at which an oil or other fluid vaporizes and ignites.

Flat Die Thread-rolling See **Thread-rolling**.

Flat Form Tools See **Dovetail Tools,** which are the most used type, though some simple flat form tools can be mounted on the normal tool posts.

Flat Lapping See **Lapping**.

Flexible Abrasive Contact Wheels Wheels for use in **abrasive belt grinding** (q.v.), and made of compressed canvas, rubber-coated canvas, solid-section canvas or buff section canvas, which are respectively of nine densities from very hard to very soft; medium-hard; soft or medium-hard; and soft. Their function, also respectively, is grinding and polishing of medium range; form polishing; polishing; and fine polishing. The compressed canvas wheel is tough and long-lasting; the rubber-coated canvas wheel forms well and also removes a good deal of metal; the solid-section canvas wheel is inexpensive and has a uniform face density; the buff-section canvas wheel is also inexpensive, but is specifically applied to fine finishing and polishing. The solid-section wheel will produce all types of polish with uniform appearance and without producing an abrasive pattern. The buff-section canvas wheel is given greater or less width by adding or taking away its sections.

Floating Reamers Sometimes termed **Floating Blade Reamers.** These are reaming (q.v.) tools whose cutting edges can be taken out and replaced or adjusted to take up wear or, in some instances, to suit variations in the diameters of the holes. Usually an adjusting screw holds the blades in a slot in the bar, and a lockscrew regulates the degree of float together with an adjusting screw in the toolholder. Tolerances of up to 0·0005 in. are possible by these means, and the tools will also correct faults in the machine or holder. Chips are rapidly cleared away and reaming is often done more speedily and economically than when solid cutters are used, the holder being re-used for other tools as required, and not needing to be taken from the machine. The work is usually rotated, not the tool. The term "floating" indicates that the reamer is not held inflexibly, but is able to move within limits in one or any direction.

Fluted Reamers See **Reamers.** Reaming tools provided with lengthwise flutes so that their cutting is done by the tool sides. The flutes are of three main types: chucking, for heavy work; straight flank; and concave flank for a large chip area. These flutes give greater chip space, and are employed where irregular spacing is desired because their lands may be readily maintained at uniform width.

Flutes Channels or grooves in tools for machining, giving the chips ready clearance, as in twist drills, reamers, taps or gear-cutting tools. The flutes may be helical or straight according to the function of the tool.

Fly Cutter A single-point cutting tool used in **milling** (q.v.) and **trepanning** (q.v.), usually of carbon or high speed steel, of round or square section, and held by special workholding devices. It is applied to a wide range of operations and is easy to manufacture, grind and control, while it may be inexpensively produced in a large number of forms. It is most economical when the number of parts to be machined is small, and has the advantage that one tool of the required shape can be produced at relatively little cost.

Fly Tools Milling tools fed at a tangent to machine gears in small numbers, or even a single gear, without the necessity of using a hob

FOLLOWING STEADY

(q.v.). The tool is made up of a finishing-hob tooth held by a screw or clamp in the arbour of the machine and fed across the blank in the same way as a tangential hob, but using a much lower feed rate. The teeth of the fly tool may be involute, straight-sided or have a special form. The cutter is inexpensive, easy to set up and regrind, suitable for many operations and with a wide range of possible forms.

Following or **Follower Steady** A steady (q.v.) secured to the rear of a lathe side rest saddle and while barely touching the workpiece just forward of the tool, travelling with the slide rest. Another term is **Back Steady Rest**. The follower steady has two adjustable jaws and supports the work to be machined.

Foot Drill An old-fashioned foot-operated drilling machine, usually of small size, and driven by a treadle.

Foot Lathe An out-of-date small lathe driven by treadle and crank worked by the foot. It is mostly found in model engineers' workshops.

Force Fit The forcing of a piece into a hole not large enough to accept it, using some mechanism for the purpose.

Form-grinding A grinding operation which involves the passing of a wheel of specific contour through a tooth space and producing the desired root depth on the left side of one tooth and the right side of its neighbour simultaneously. Thus, the desired contour is given to the work. Such wheels are used for cutting gears, form cutting tools, taps, some reamers and also contoured tools for lathes, screw machines, etc. (See **Grinding**.) Form grinding can also be done by **electrochemical grinding** (q.v.) and **electrical discharge grinding** (q.v.).

Cylindrical grinding machines are sometimes used to produce intricate contours in rough work, while surface grinding machines with a horizontal spindle are able to grind contours of virtually all kinds. Whether or not to form-grind a component depends largely on the numbers involved and the need for economy.

The wheels may be either crush-trued or diamond-trued, both methods having their advantages.

Form-milling See also **Milling**. A milling operation carried out with cutting teeth formed with a radial or curved profile to obtain the required degree of relief. They can be re-sharpened by grinding without modification of the tooth form. The concave and convex milling cutters are typical examples, producing either concave slots or convex edges. Only the front faces have to be ground, and the cutters reproduce the same profile within narrow limits until worn out. The tools are used in milling out the teeth of gears, and on aluminium alloys and magnesium alloys, steels, etc. They may be carbide-tipped or of tool steel.

Form-milling is more expensive than planing but removes metal more efficiently in a specified time. It is less efficient than grinding when the quantity of material to be taken off is small and greater accuracy in dimensions and surface finish is required, but in many instances both operations are combined, though grinding alone is preferable for such work.

Form-tools Tools of pre-established form fed into a rotating workpiece at 90 deg. to the axis of the work, to develop the desired work form without any longitudinal tool motion. They were developed for mass production, and the contours generated often comprise intricate curves with narrow projections and recesses, etc. They are manufactured to extremely fine limits, and produce parts with curved forms or those in which curves, recesses, projections and angles are combined, without the necessity of using a large number of single tools with their respective motions and variations. Moreover, all the parts are identical and no skilled labour is required to produce them.

Formate Cutting A technique for machining spiral bevel and hypoid gears which have to be both roughed or single-cycle finished and have pitch diameters up to 33 in. The technique is non-generative, and one machine may often be employed for both operations, but better results are obtained by two machines. In single-cycle finishing a single tooth-space is machined with a single rotation of the cutting tool, which consists of circular blades held in a circular cutting tool somewhat akin to a face-milling cutter. The blades are of varying length and width, each being longer and wider than its successor, and a space between the first and final blades permits indexing (q.v.) as each tooth-space is finished.

Free-cutting Steels Steels machinable without difficulty, and not producing long, tenacious chips liable to build up on the cutting tools, thus hampering them. They are especially important when automatic machining in screw machines and turret lathes is employed, and often have special alloying elements added to their composition for this purpose, among them being bismuth, lead, selenium or sulphur, the actual content being carefully controlled. Sometimes they are known as **Free-machining Steels.**

Friction Bandsawing A means of cutting ferrous metals by frictional heat rather than by the cutting action of teeth. The saws are run at high speeds (6,000–15,000 surface ft./min.) and the heat of friction softens or liquefies the metal before the carbon steel saw teeth get to work. The teeth have then only to take away the metal

FRICTION SAWING

from the cut. The saw blade does not lose temper or become too hot since its contact with the metal is small at such high speeds (vastly in excess of those of ordinary bandsawing), and the frictional heat generated is rapidly dispersed. The saws are of greater thickness than those of ordinary metal-cutting bandsaws and are of wider set, with teeth capable of supporting severe shear loads, and having gullets of special form to ensure an even flow of chips.

They are mostly used for ferrous metals of greater hardness than Rockwell C_{42} or which are likely to harden on the surface after a short period of work. They also cut thin sheets with little or no work-distortion, and achieve both straight cuts and intricate forms with high rates of sawing and extensive service-life of the saw, which fails only after fracture from continual flexing. They are inexpensive for material up to 1 in. thick, having low heat conductivity combined with a considerable temperature range for softening. The hardness of the metal is not of great importance.

Cast irons are not suitable for this operation, nor are some of the non-ferrous alloys, such as those of copper and aluminium.

Friction Sawing A method of cutting metals by means of frictional heat, but using a circular disc rather than a bandsaw. The disc is notched and runs at a peripheral speed of 25,000–30,000 ft./min. and *burns* rather than cuts its way through the work. A better name for the tools is **friction discs.** Made in various sizes and gauges from a special steel they are specially useful for quick, rough cuts on cold material. The chief standard sizes are 46 in. dia. × $\frac{5}{16}$ in. thick, or 63 in. dia. × $\frac{5}{16}$ in. thick. To cut sections of different thicknesses, it is not necessary to change the blade. Water sprayed on the heated blade in use may cause cracking on the periphery, so to prevent this the amount sprayed is moderate. The disc is regularly turned or ground true and correct tension maintained. A high-speed friction sawing machine for cutting rolled steel sections, joists, channels, angles, etc., is massively built to prevent vibration. It has a powerful motor and can accept large overloads, the spindle, specially large in diameter, revolving in roller bearings. The capacity on joists is up to 24 in. × $7\frac{1}{2}$ in. The d.c. machine takes a blade of 46 in. dia. and weighs about $4\frac{1}{2}$ tons net.

Front Clearance Angle See **Tool Angles.**

G

Gang Cutters These are **Milling Cutters** (q.v.) mounted on the same arbour.

Gang-drilling Machines Machines with a minimum of two spindles in line on the same base and work-table. They are manually or power operated and can be separately set to give the desired speed and depth of cut. The number of spindles is governed by the number of different operations to be performed. They are mostly used for quantity production of holes in a number of identical components.

Gang Milling A method of **Milling** (q.v.) in which two or more milling cutters, either identical or different in dimensions, are placed on the same arbour, with the work surfaces either neighbouring or some distance apart. The technique is usually employed to give a number of different steps in the work at one and the same time, or to give sections of identical thickness from the stock material.

Gang-planing Sometimes termed **Tandem Planing,** this is a method of planing (q.v.) in which the components are firmly placed end-to-end in a row to ensure that the planing tool cuts them continuously—that is to say, a "string" of parts is planed. On the other hand they may be about 6–8 in. apart to give a cut that although discontinuous does not injure the tool by the resultant vibration since the components are fairly close together.

Gang-tapping A technique of **tapping** (q.v.) designed for a high rate of output, but also for producing small numbers of parts owing to the facility with which the machines can be set up. It allows of in-line operations such as drilling, reaming and tapping.

Gap Bed Lathe A lathe bed having that portion close to the faceplate or chuck below the level of the bed on which the carriage slides. This enables work of somewhat greater diameter than that determined by the height of the centres above the bed sliding face to be revolved. See Fig. 14.

Gap Gauge A **gauge** (q.v.) used for measuring the gap or distance between some part of a workpiece or other product.

Garnet A medium soft vitreous abrasive with a grit size of 600–800 microns, used mostly for finish-lapping of brass and bronze or in grit size of 10 microns for polishing these non-ferrous alloys.

Gas Laser See **Laser Machining.**

GAUGE BLOCKS

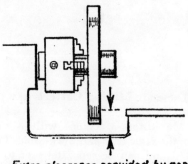

Extra clearance provided by gap

Fig. 14   Gap bed lathe

Gauge Blocks Hardened tool steel blocks made to a high degree of accuracy and used in machining for testing parts to such fine limits of measurement that standard shop gauges will not serve. They are in boxed sets, each one of the set being of specific thickness and length and marked with its correct dimension, representing the distance between two opposing parallel surfaces. For all practical purposes they are exact to a few millionths of an inch. By combining the blocks a wide range of precise dimensions can be had. Though the number of blocks in a set is variable 81 is the most common. The blocks do not advance consistently in size, some advancing by 1/10,000th in., others by 1/100th, others again by 1/25th, and a final group by inches. Small individual sets for large sizes are made containing blocks up to 20 in., and some of small sizes contain only blocks as thin as 1/100th in. The gauge blocks are laid one on the other, and worked or rubbed round until complete contact is obtained, sliding one slightly off the other, then pressing them vigorously together—this is termed "wringing together". Precision gauge blocks are often used to test **Master Gauges** (q.v.) and ordinary gauges. They are sometimes built-up in adjustable holders.

Gauges Instruments of fixed and exact dimensions indicating quickly and accurately whether a component has been machined to the limits of size specified. Some gauges are adjustable, but in mass-production non-adjustable gauges are more usual. There are numerous types, including internal and external gauges, internal limit gauges, caliper gauges, standard taper gauges, screw gauges, Wickman gauges (q.v.), position gauges, step gauges, thickness and radius gauges, depth gauges, etc. Gauges have to be tested themselves from

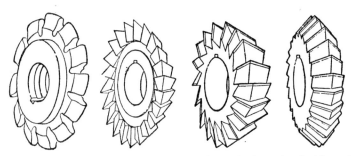

Figs. 15–18   Gear cutters: l. to r., *involute, single-angle, double-angle, equal-angle*

time to time and this is done by a **Master Gauge** (q.v.). Some of these gauges are briefly described under their own headings.
Gear Cutters Tools employed to produce gears by giving the teeth the necessary contours and dimensions, or by forming the correct tooth contours by relative movements of cutter and gear, as when a straight-sided tool generates the desired tooth contours by means of the generating motions. Sometimes templates or master formers give the tool its true path and so develop the requisite tooth form, as when gears of specially large size are machine-cut.

Involute gear cutters (see Fig. 15) are used in the milling machine to form the spaces between the gear teeth and produce their own form. There are normally 8 of these cutters to a set, capable of cutting gear wheels of 12 or more teeth. Sets of 15 of these cutters are used when the gears have teeth closer to the theoretical involute form. There are also **single-angle cutters** (see Fig. 16) and **double-angle cutters** (see Fig. 17) with side teeth. The single-angle cutters may be right- or left-hand and their corners are filleted or rounded off to give the tools greater service life. Any angle of cutter can be obtained. The double-angle cutters have their two faces unequally inclined, one angle usually, though not necessarily, 12 deg., since in cutting helical work this angle prevents interference on the perpendicular cut. They are used mostly for cutting helical gears, and are made left- or right-hand. **Equal-angle cutters** (see Fig. 18) are double-angle cutters with equal inclination of the teeth and no "hand", being made to any angle, but normally 45, 60 or 90 deg.
Gear Cutting Methods of producing the teeth of gears. With the exception of bevel gears, the operations involved, which should be referred to under their individual headings, include **milling, broaching, shear cutting, hobbing, shaping** and **rack machin-**

GEAR GENERATION

ing. The operations in each case necessitate some method of holding the work in proper relation to the cutting tool, and once each tooth space has been cut, the blank is usually indexed (q.v.) to the following position. The tools used may be either form tools (q.v.) or circular revolving cutting tools. The various techniques are extremely complex, and modern needs have necessitated a considerable variety of machine tools to produce the numerous forms of gears required.

Bevel gears are machined in machines of special design using tools which are also of special design, but these machines, though not specifically milling, shaping or other machine tools, are in general similar to them in the mode of movement of their cutters. They may be either generating or non-generating. Characteristic techniques dealt with under their own headings include **milling, template machining, formate cutting, helixform cutting, cyclex cutting, face-mill cutting, interlocking gear cutting, revacycle cutting** and **planing generation.** Many of the gears can sometimes be machined by more than one of these processes.

Gear Generation See **Gear Grinding.**

Gear Grinding A means of producing hardened and heat-treated low carbon steel gears of straight spur type to give them a better finish. The advantages are: correction of distortion after heat-treatment; quieter running of the gears; high output at comparatively small cost; great reduction in gearbox costs; no limit on the type of steel to be ground; gears running at high peripheral speeds with long service life; a high safety factor if the gear is correctly designed; accuracy, uniformity and smoothness of tooth finish in small gears for special work.

The teeth are ground either by the generating method or by the formed wheel method. The **generated tooth** is finished by the flat or angular side of an abrasive wheel, which generates the tooth form as it grinds. The formed wheel finish-grinds the entire tooth-space, the wheel passing through the spaces until the true size is obtained. This gives extremely accurate gears, is economical and copes with a considerable range of gear sizes. As much as 0·03 in. can be removed from each side of the tooth space. Costs are low and tooth-roughing speedy, allowances being made for excessive distortion.

In a third method two tooth surfaces are ground simultaneously, two wheels working in different tooth spaces, using their flat sides.

Some teeth are formed entirely by grinding, but these are mostly of fine pitch when the quantity of material to be ground off is small. The generating method of grinding is invariably employed for helical bevel and hypoid gears. Cost is usually the deciding factor in machining or grinding for gear cutting. Aluminium oxide abrasive-wheels are preferred for hardened and case-hardened steels, but for nitrided steels silicon carbide wheels are used on occasion. The grit sizes of grinding wheels for gears range from 46–100, but as low as 400 may be used for gears of diametral pitch of 200 or thereabouts. Vitrified bond wheels (q.v.) are usually preferred, but some resinoid bond wheels of cup form are used in generating bevel gears. The grinding speeds are usually from 5,500–6,500 surface ft./min., but higher and slower speeds may be used for special work. In-feed is determined by the gear pitch, and ranges from 0·0015 to 0·0005 in./pass.

Gear Hobbing See **Hobbing**.

Gear Shaping Generating the teeth of gears of herringbone or helical type (helix angles up to 45 deg.). Two cutters are used at the same time for herringbone gears in a shaping machine, one cutter being used for each helix. The tools reciprocate, one machining towards the work centre, the other directed towards the same point when the movement is reversed, but coming from the opposite direction. Both tools revolve at a slow speed and generate the teeth in the manner of a normal shaping machine. (See **Shaping**.)

In shaping helical gears an extra rotary movement is given to the reciprocating spindle holding the cutting tools, and a helix guide regulates the rotation. The guide lead is identical with the tool lead.

Gear Stocking Cutter A form of milling cutter applied to the cutting of gear teeth in a roughing stage, so that the tool eventually used for finishing may have less wear imposed upon it and provide a gear more accurate to finished dimensions.

Glazed Grinding Wheels Grinding wheels dulled by the attrition and wear of the grains, so that there is not enough breakage of the grains to give the desirable sharp cutting edges. In consequence the wheel has a smooth, shiny surface. Glazing of the wheels is liable to generate excessive heat in grinding.

Grain Size In grinding wheels, see **Grit Size**.

Graphite Electrodes Electrodes used in **electrical discharge machining, electrical machining** and **electrochemical machining.** They are machined and ground in the normal way and finished by hand. The dust created must be removed or lessened by paraffin applied during the operations. These electrodes are less costly than metal electrodes, have long service life and can be machined without difficulty. Sometimes the graphite is used as a tip on a punch made of tool steel, the tip being applied by soldering or organic cement, or even by some method of mechanical attachment. The electrodes are sometimes used for the electrical machining of stainless steel, having a longer service life than electrodes of copper-tungsten alloy type.

Graphite Grinding Wheels Wheels used in **electrical discharge grinding** (q.v.) and running at 100–600 surface ft./min. They are more rapidly and economically dressed than diamond wheels when carbide tools are being ground by this technique. They are also used in **electrochemical discharge grinding** (q.v.) and here their flat faces are dressed by a cemented carbide or high speed steel tool. The service life of the wheels depends on the type of graphite, the operating voltage and the power supply. In profile- or form-grinding the wheel is sometimes formed in position by plunge-cutting (q.v.), using a "master" of high speed tool steel, and can be similarly dressed. In this form of grinding the wheel speed is from 4,000–6,000 surface ft./min., which range should not be exceeded.

Graphite Scrapers Graphite pieces having a thickness of about $\frac{3}{8}$ in. pressed against a rotating grinding wheel in **electrochemical discharge grinding** (q.v.) to form multiple-form contours without flat vertical or horizontal surfaces. The pressure against the wheel is of the order of 10 lb./sq. in. and the scrapers are profile-ground. Graphite scrapers can also be used to grind intricate contours with both curves and flat vertical or horizontal surfaces. The scraper is hollow and air under compression is forced through the hollow to regulate the distribution of electrolyte, particularly where the surfaces are horizontal or flat. Here the pressure against the wheel is about 15–30 lb./sq. in., the air pressure being 40–60 lb./sq. in. and the wheel being set 0·002–0·007 in. apart from the scraper.

Grinding The removal of material from an object by the action of an abrasive substance: a grinding wheel with its surface suitably coated is frequently used. There are various forms of grinding, namely cylindrical, internal, surface, disc, centreless, thread, roll, tool and gear, while some small holes are ground with diamond mandrels. The refined grinding techniques include lapping, honing and superfinishing. Not all grinding is done by wheels. Sometimes abrasive belts are used, and there are in addition polishing and buffing operations which are also examples of surface-improvement by abrasive action.

The abrasives used in grinding may be natural or artificial. The natural include emery and sandstone, corundum, garnet, pumice, diamond dust, kieselguhr, Tripoli powder, rottenstone, etc. Artificial abrasives include diamond, steel, emery, jeweller's rouge, crocus, whiting, manganese dioxide, bath brick, calcined china or pipe clay, aluminium oxide, silicon carbide, etc., aluminium oxide and silicon carbide being the most important. The oxide is chiefly used for cutting high tensile materials, the carbide for non-ferrous alloys and cast iron of low tensile strength, but there are exceptions in both instances.

Grinding may be used to either give suitable precision to form and dimensions or bring rough forgings or castings to proper dimensions without the need of cutting tools. The machines mostly used include cylindrical grinders, internal grinders, centreless grinders and flat-surface grinders. Wheel speeds range from 5,000–6,500 surface ft./min. or may go up to 15,000 surface ft./min. The speed is mainly governed by the kind of wheel and the material to be ground.

Grinding Burn A change in the hue of a metallic surface is sometimes produced by the temperature to which the work-material is heated by friction. It is caused by a film of oxide, which may be lemon-coloured, brown or blue in varying shades, and usually the steels in particular show this effect. Sometimes the tint is harmless but in other instances, with heat-sensitive materials, it may produce a disadvantageous microstructure of the metal.

Grinding Fluids As with machining, grinding operations are liable to generate frictional heat, which may injure the wheel surface and spoil the surface finish. Hence, it is essential in most instances to use a **cutting fluid** (q.v.). This prevents distortion of the work and removes particles of abrasive dust, while also producing a better surface finish. Too much coolant is better than too little, and it usually contains ingredients designed to protect both machine and work from rust. Soda and soluble oil in the proportion of 1 lb. soda and 1 pint oil to 10 gall. water are often used. The coolant is directed on to that area of the work coming into contact with the wheel. The flow is wide enough to cover the wheel face completely, so preventing feed markings, i.e. marks left on the work-surface by the wheel-feeding motion. This involves using a special pipe nozzle and a baffle or deflecting-plate to widen the coolant-spread and direct it accurately. The volume of coolant used depends on wheel diameter and thickness, the work material and the speed.

Coolant pumps are normally provided with the grinding machine. Where dust and chips are plentiful, as in powerful vertical spindle machines using cylinder wheels, as much as 40 gall./min. is directed against the wheels, both inside and outside. In grinding threads, oil is the most suitable coolant and also lubricates, so preventing the wheel from becoming loaded, while it facilitates abrasive cutting and lessens failures. Finer grit abrasives can be used when the coolant is oil than when a water solution is used, and oil also aids in re-

GRINDING WHEELS

taining the surface-ground properties. Nozzle design for the coolant is important, especially when the wheels run at high speed. Pressure should ensure correct volume for the operation to prevent dispersal of the stream by air-flow round the wheel. The oil tanks are regularly cleaned and the oil filtered and separated if necessary to eliminate metallic particles that might injure the finished worksurface and clog the wheel. An oil cooler is of value also.

In **abrasive belt grinding** water is sometimes, after the belt has passed over the top pulley, sprayed on to it to wash away particles, so giving freer cutting, better work surface finish and longer belt service life. Centrifugal force discharges cuttings and dust from the belt surface as it passes round the bottom pulley.

Dry grinding is often used for lighter grinding machines with horizontal spindles, especially for grinding high speed steels with a soft grade disc wheel.

Grinding Wheels Wheels of the correct abrasive, grain size or grit size, hardness and structure, for the grinding work they have to do. Most steel cutting tools are ground with aluminium oxide wheels, but for some exceptionally hard materials, such as tungsten carbide, silicon carbide wheels are used. The greater the areas in contact, the coarser the grit. Wheels of large diameter having a greater arc of contact with the work require a softer grade, as do hard steels. Finishing work calls for a finer grit, and higher work-speed for a harder wheel.

Standard plain grinding wheels are obtainable and have either a 60 deg. or 45 deg. profile angle. Other forms and types of wheel are standardized. A British Standard system uses letters of the alphabet to indicate the hardness of every type of wheel bond, starting with the softest and finishing with the hardest. (This system is not applicable to diamond wheels, oil stones, scythe stones, etc.) The markings are classified into abrasive type, grain size, grade and type of bond. Grade is shown by letters from A, the softest, to Z, the hardest. Wheel structure is shown by a number 1, the densest, and up to 15, the most open. The bond is also shown by letters, thus V is vitrified, R is rubber, S is silicate and B resinoid, while shellac is denoted by E.

The American Standard B.5.17 is somewhat different. If we take for example a grinding wheel classified as 51–A–36–L–5–V–23, the origin of the classification is as follows:

51: this is the maker's symbol for the precise type of abrasive used, and the inclusion of this number is in fact optional.

A: This indicates the type of abrasive, Aluminium Oxide being identified with an A and Silicon Carbide with an S.

GRIT SIZE

36: This is the grain-size, grain-sizes being as follows:

Coarse	Medium	Fine	Very Fine
10	30	70	220
12	36	80	240
14	46	90	280
16	54	100	320
20	60	120	400
24		150	500
		180	600

L: The grade. Grades run from A (soft) through to Z (hard).
5: This is an indication of the structure, and its inclusion is optional. The figure derives from:

Dense	to	Open
1		9
2		10
3		11
4		12
5		13
6		14
7		15
8		etc.

V: This symbol indicates the type of bond used. V=vitrified, S=silicate, R=rubber, RF=reinforced rubber, B=resinoid, BF=reinforced resinoid, E=shellac, and O=oxychloride.

23: The maker's private identification mark. Its use is optional.

Grit Size The approximate size of the abrasive grit particles used in building up a grinding wheel, representing their fineness or coarseness. When the abrasive has been crushed, it is classified by pssing it over a standard screen or sieve with holes of known size, so many to the linear inch. The range is from 8–240, though the last size is used when a "flour" only is needed, obtained by other methods than screening. The sizes are shown by numbers corresponding to the number of holes or meshes in the screen and are shown in Table VII (page 98).

A wheel composed of uniformly-sized grains is termed a straight grain wheel. Combination grain wheels, as the blended grit wheels are called, usually require skill in their blending and follow a definite pattern. Table VIII (page 99) is a diamond-wheel-size chart supplied by a large manufacturer. There is a slight difference when diamond tools are used for dressing and truing abrasive wheels and

TABLE VII
Grit Sizes of Typical Grinding Wheels

Abrasive	Coarse	Grain Size		Very Fine
		Medium	Fine	
Alundum				
A	10	30	70	220
15A	12	36	80	240
19A	14	36+	90	280
35A	16	46	100	320
38A	20	54	120	400
57A	24	60	150	500
Crystolon				
37C			180	600
39C				

for lapping. Any grit size 60 and finer is used, the coarser grits removing material faster, but giving a less attractive finish.

Grooving A machining operation in which grooves are machined in aluminium alloy automobile pistons, etc. The tool used is a parting tool (q.v.), set as near to the toolpost as possible and given ample side clearance. The roughing speed used is 300 ft./min. with ordinary tools, or 500 ft./min. with specially-ground tools. The top rake is 15 deg., and the front clearance about 5 deg. The term is also used for the recessing of wooden boards on the edges to take the corresponding tongued boards, the work being done by rotating cutters or a small circular saw with wide teeth and of somewhat greater thickness.

Ground Thread Taps Taps applied to produce concentric threads in holes. They have a pilot (q.v.), guided by the hole, but sometimes a bush is used instead. They are not standard tools, but are made specially when required. See **Tapping**.

Guide Any device enabling a piece of mechanism such as a crosshead to travel compulsorily along a single path of whatever form required.

A **Guide Bush** is that piece receiving the guide post. The **Guide Post** is that part of, for example, a die set that links the top and bottom of the part; the **Guide Holes** are the apertures in which the posts and bushes are placed.

Guide Screw The leading screw of a self-acting lathe, employed as a means of traversing the slide rest and starting screw threads.

TABLE VIII
Diamond Wheel Size Chart

F Rating	Wheel Dia. (in.)	Wheel Face (in.)	Grit	Grade	Bond	Abrasive	Carat Value
1	up to 6	1	40–80	Soft	Vit. & Sil.	Al. Oxide	
2	6–12	2	40–80	Med.	Shellac	Al. Oxide	
3	12–16	2	90 and finer	Med.	Shellac	Al. Oxide	
4	16–20	3	90 and finer	Med.	Shellac	Al. Oxide	
5	20–24	3	90 and finer	Med.	Shellac	Al. Oxide	
6	24–30	4	12–36	Hard	Rubber	Sil. Carbide	1
7	30–36	4	12–36	Hard	Rubber	Sil. Carbide	1
8	36–40	5	12–36	Hard	Rubber	Sil. Carbide	1
10	36–40	6	12–36	Hard	Rubber	Sil. Carbide	2
12	36–40	7	12–36	Very Hard	Rubber	Sil. Carbide	2
14	36–40	8	12–36	Very Hard	Rubber	Sil. Carbide	2
16	36–40	9	12–36	Very Hard	Rubber	Sil. Carbide	2
18	36–40	10	12–36	Very Hard	Rubber	Sil. Carbide	3

Guillotine A machine for shearing or trimming off metal when continuously travelling as in cutting corrugated or profiled metal strip.

Gullets The deep spaces between the teeth of saws, broaches, etc., allowing the chips to escape. In broaches, large gullets take chips in coils, and are often deepened to take a larger volume of chips, but if this would weaken the tool, a longer gullet is provided with a greater tooth pitch, the broach length and stroke being increased. In resharpening a broach the hook angle must not be changed nor the form of the gullet. Some special saws have deeper teeth and hollow tooth roots, inclined outwards on alternate sides to enable the chippings to escape more freely. Bandsaws have a deep gullet with a smooth radius at the bottom.

Gun Drilling A type of drilling carried out on a horizontal machine tool. (See **Drilling**.) It employs a drill of special design for drilling deep blind holes within comparatively fine dimensional limits. The drills have one flute of vee form whose cutting face is ground to give two cutting angles each producing a chip of little length so that they can be readily cleared. The body of the drill is of oil-tube (q.v.) type for coolant and lubricant supply to the drill point and readier chip clearance. The centring of the drill is maintained by carbide or other wear-resistant material in the form of pads inserted opposite to the cutting face.

The machine must have a positive feed and complete rigidity and, if this is achieved, the drill produces a more accurate hole than a twist drill, and thus may reduce the number of finishing operations, such as rough and finish reaming, or even eliminate them. The drills will give a long hole, straight and round, within strict limits of size and concentricity in aluminium alloys, cast iron, heat-resisting alloys, magnesium alloys, copper alloys, stainless and other steels, and titanium alloys. Gun-drills and their equipment are, however, costly and not always economical. In drilling nickel and its alloys the drills are mostly used for deep holes up to and including 2 in. dia., or even at times up to $3\frac{1}{2}$ in. dia.

The lubricant is mostly a highly-sulphurized cutting-oil, fed through the tube in the body at a pressure of about 800 lb./sq. in. for holes $\frac{3}{16}$ in. dia., down to about 200 lb./sq. in. for holes 2 in. dia.

In gun-drilling stainless steels a carbide-tipped drill is used with special angles to suit the material, namely 30 deg. outer point angle and 20 deg. inner point angle. Some qualities of stainless steels are drilled with step-point gun drills, which are said to improve chip clearance. A few drills for this class of work are made of solid tungsten carbide, but this is unusual.

GUN REAMERS

Gun Reamers Tools for boring and finishing holes, used when the work is rotated and the tool is fixed. They are similar in design to **gun drills** (q.v.), and are supplied in $\frac{1}{8}$–2 in. dia. Some tools of this type have detachable heads for easy replacement. They all give the same advantages as gun drills, namely straightness of hole, close limits and fine finish, and run at about 200 surface ft./min., with a feed of up to 0·01 on the softer materials. The reamer flutes pilot themselves like the gun drills and run the entire length so that the chips are cleared away rearwards. The cutter has a round shank, but is cut away at a certain point to make it possible for the lubricant to flow.

Gun Taps See **Helical Point Taps.**

H

Hacksawing Sawing steel and other materials with a saw, once made solely of carbon steel, in the form of a short band with a toothed edge, each end being rounded and pierced with holes to permit use in a hacksawing machine. Today few carbon steel blades are produced, most blades being made from a 1·5–2·0 per cent tungsten steel with a small vanadium content or from high speed steel containing 14–18 per cent tungsten. They may be used in a frame for hand sawing or in a machine specially made for this type of work, to cut up aluminium alloys, copper alloys, heat-resisting alloys, magnesium alloys, titanium alloys and tool steels. Aluminium alloys are usually cut off with a power hacksaw when straight cuts are required, and this machine is preferable to a circular or band saw or an abrasive wheel. The cutting is normally at 140–160 strokes/min. with a feed of 0·015 in., the higher speed being preferred for unheat-treated cast aluminium alloys. A cutting-fluid is necessary except when the cuts are extremely light, and a fluid lubricant is best (see **Cutting Fluids**).

For *copper alloys* hacksawing speeds range from 90–150 strokes/min. according to the hardness of the alloy, with a feed-range of 0·010–0·012 in./stroke. For power hacksawing heat-resisting alloys the speed range is 20–70 strokes/min. with a feed range of 0·003–0·006 in./stroke for materials of hardnesses from Brinell 140–320.

HAND-DRILLING

Power hacksaws are much used in the sawing of magnesium alloys and are run at speeds up to 160 strokes/min. The blades are given a tooth pitch of 2–6, a tooth set of 0·015–0·030 in., and a clearance angle of 20–30 deg.

Not all *nickel alloys* lend themselves to power or hand hacksawing. Those that do are power-sawn with heavy-duty high speed steel blades having a tooth pitch of 6–10 and a raker set (q.v.), but this applies to solid bars only. Tubes require 14–18 pitch, or 24–32 pitch for thin walls, or not less than 18 pitch for less than $\frac{1}{16}$ in. wall, when hand sawn.

Titanium alloys are power-hacksawn at a speed of 25–180 strokes/min. and feed of 0·003–0·009 in./stroke, for alloys of Brinell 110–400. Some refractory metals can be hacksawn, such as niobium (columbium), tantalum and molybdenum, at speeds ranging from 40–90 strokes/min. and feed 0·006 in./stroke, for metals of Brinell 170–290 hardness. Tungsten should not be hacksawn.

Tool steels are hacksawn with high speed steel blades at speeds from 45–140 strokes/min. and feeds of 0·003–0·006 in./strokes, for materials of Brinell hardness 100–375, according to the type of steel.

The blades in power- or hand-hacksawing are reciprocated. In modern machines the blade cuts on the draw-stroke and is taken out of the cut on the return-stroke to prevent the blade from being dulled. An adjustable weight is provided by a spring and dashpot and the machine carries a pressure indicator. Coolant is supplied to the blade when cutting by pump from a tank, and is drained back, filtered and re-used. Many machines are now electrical in operation and of high power. These use blades from $1\frac{1}{4}$–$2\frac{1}{2}$ in. wide, 16 gauge or thicker.

The work is held in a work-table in a powerful screw vice and automatic control of the lengths cut is combined with automatic stop motion on the electrical machine so that it can be arrested at any depth of cut. On completion of the cut the blade is withdrawn, moves the prescribed distance along the side rods or bars, then continues to cut. It is also possible to fix the work directly to the work-table.

Specially heavy hacksawing machines vertically saw joists measuring 20 in. $\times$ $7\frac{1}{2}$ in. so that they can be mitred. Feed governors are provided on some machines.

In many works cut-off **bandsawing** (q.v.) has replaced hacksawing, especially for short runs.

Hand-drilling Drilling may be done by hand in a **drill press** (q.v.), or a hand ratchet drill, bow or hand-brace.

Hand-tapping Tapping carried out on a hand-operated machine when only a few components are required and the use of powered machinery would be uneconomical. It is quite possible to produce precision and high quality threads in this way. Despite this, most taps are now used in powered machines, but for some work the old name of "hand taps" is still used. They have straight or spiral flutes, but the number of flutes, from 2–4, depends on the material being tapped. These tools are rarely employed for tapping blind holes. Hand-tapping of stainless steels is advantageous when only small quantities are required. The tap flutes are usually milled out by a milling cutter.

Headstock The mechanism that rotates the work in a lathe. It is secured to the lathe bed, normally leftwards of the operator, and contains the main driving spindle for the work and gearing to control the rate of revolution. It may have a cone-shaped centre to support the work and either a face-plate, fixed to the spindle, or a similarly fixed suitable attachment for driving the work. Alternatively a chuck may be included for holding the work.

On the operator's right is the **tailstock** (q.v.).

Heavy Cut The operation of cutting metal in which the tool is made to produce thick, wide cuttings, as in roughing. The term is, however, a loose shop expression, variable according to the class of work, the operator and the size and composition of the workpiece.

Heel-cutting Tap A special type of tap designed to provide a satisfactory finish combined with precision. The tool is given a heel rake angle of 3–5 deg., but without relief, to prevent clogging of the hole by chips.

Helical Mills See **Milling Cutters.**

Helical Point Taps Straight-fluted taps whose flutes are angular and left-hand close to the point to force the chips forward before the tool. They are principally recommended for tapping completed holes, but can also be used for blind holes when given a plug chamfer, and they will then provide adequate clearance to accept the chips. The flutes are not so deep as in ordinary hand taps, since the chip disposal required along them is less, and because of this the body of the tap is more robust. The taps have a shearing action when cutting so that the threads produced are well finished. Moreover the minimization of chip travel along the flutes facilitates the supply of cutting fluid to the tool edges. They are sometimes known as **Spiral-point Taps.**

Helically-cut Resistors Specially designed, a tool for the manufacture of helically-cut resistors. Batches are produced with a spread as close as ± 5 per cent from the nominal of 1,000 ohms. They are cut in a lathe-type device operating with a grinding wheel to provide a helical cutting machine. (Br. Pat. 1,148,261.)

Helix Angles See **Tool Angles.** The inclination of a helical curve in relation to its axis. When the periphery of a spiral gear is divided by its lead (the distance a tool would move forward axially in a full revolution), the quotient is equal to the tangent of the helix angle relative to its axis.

Helixform Cutting A means of machining helical bevel and hypoid gears by a non-generative process. A single turn of the cutter used completes each side of a tooth space, the cutter both reciprocating and rotating, so that the tips of the cutting blades run tangentially to the gear root plane. The cutting edge being in a straight line, the tool cuts a true helical surface. In other respects the technique resembles **Formate Cutting** (q.v.) but has the advantage that the gear machine is reciprocally related to or joined with the mating pinion so that the consequent contact pattern shows virtually no bias. Also the gears are more rapidly machined at less cost.

High-helix Drills See **Twist Drills.** Drills having a helix angle of about 40 deg. designed for the cutting of special materials such as aluminium, copper, fibre, magnesium, marble, slate and wood, when holes of considerable depth are required.

High Speed Steels Steels for cutting-tools used on metals that cannot be economically machined with carbon or ordinary alloy tool steels. They contain varying amounts of hard metallic elements such as tungsten, chromium, vanadium, molybdenum, cobalt, titanium, etc., with a steel base. They are often known by their composition, e.g. "cobalt high speed steels"; or by numerals such as "18-4-1", indicating 18 per cent tungsten, 4 per cent chromium and 1-2 per cent vanadium.

Hob Taps Taps used for hobbing out the thread of spring screw dies to give greater back clearance. Taper hobs are favoured for this work.

Hobbing In machining, a technique for cutting the threads of worm wheels, dies or chasers, using a tool known as a hob or a master tap in a lathe. It also includes the machining of gear teeth in specially-constructed machine tools. Gears are cut in hobbing machines by the rotation of both hob and work at the same time.

Many special shapes can be hobbed, but the ordinary machines for this work cannot be applied to bevel and internal gears. Work and hob rotate in a constant relation, the hob being fed across the width of the work face. The actual tool or hob is a worm wheel having contoured teeth with a relief angle, so that as these enter the work one after the other, each takes up a somewhat different location, the teeth of the gear being steadily generated as the cuts proceed.

HOBBING

Hobbing is lower in cost than either broaching or a shear-cutting action, and is therefore mostly applied to small lots, but it is speedy and gives precise dimensions, and can consequently be used for large numbers if required. Hobs can be obtained capable of machining spur and helical gears of more than 100 in. external diameter, but for this purpose the diametral pitch is less than 1, as gears above this pitch are by no means easily machined. Hobs are supplied either ground or unground according to the degree of accuracy. Ground jobs are resharpened with care to ensure maintenance of dimensional accuracy and tooth contour.

Hobs will machine the teeth of herringbone gears if the cutter is free to run out between right- and left-hand helixes, which means that the space between the teeth must be sufficiently wide to take the tool. Gears of this type are hobbed up to 220 in. For cutting worm wheels hobs have the teeth at the entering end bevelled off to minimize the load. If a higher degree of surface finish is required on a worm than the ordinary hob will give, the flutes are increased in number, as this diminishes the marks of the feed.

Spur gears are mostly hobbed because of the extremely accurate cutting over a considerable dimensional range, the ability to cut either small or large quantities, the comparatively inexpensive cutting, and the ability to machine materials more than usually hard. The form of the gear may restrict the applicability of the technique, as when part of the work diameter exceeds that of the gear root and the hob teeth have to cut near to this area. In such instances the hob must travel almost half the hob diameter over the axial distance separating the sections. In addition, between gear and any projection such as a flange, a clearance of about half the gear diameter must be given together with extra clearance to provide for the angle of the hob thread.

Helical gears are machined by hobbing much as are spur gears, but the rotation of the gear blank is advanced or retarded in relation to the hob rotation, the feed being also related to blank and hob. If the hob has a number of flutes or gashes, the blank rotation is advanced, but for a single-fluted hob it is retarded, advance and retardation depending in amount on the angle of the helix. Where considerable output is required fixtures for hobbing a multiplicity of blanks in a single loading of the machine are usual.

Hobbing is specially advantageous for machining the teeth of hard steel hobs up to Rockwell C48, but for this purpose there has to be no play in the machine, both tool and blank being securely held in position. Wear of the hob is greater as the work material is harder.

HOBBING MACHINES

Hobbing Machines Machine tools that cause a gear blank and a hob to rotate together, the relation between them being governed by the number of teeth in the gear and the number of flutes on the hob, i.e. from 1 upwards. The hob rotation enables following teeth to take up locations like those of a rack (q.v.) in mesh with the rotating blank and travelling tangentially. As the hob rotates, the machine slide carrying it feeds in a direction parallel to the blank axis, and feed is continued until all the teeth have been properly cut. This is achieved by a single travel of the hob, all the teeth being cut together.

The hob has to be set to the proper angle and the machine properly geared to ensure that blank and tool rotate at the correct speed ratio, using change gears for this purpose. The hob-setting angle is governed by the lead of the hob flutes and the hob diameter, and is usually marked on the tool itself when obtained.

Hog-nose Drill An American shop term for a type of drill possessing great rigidity and used in coring out holes. It resembles a boring tool (q.v.).

Hollow Drills See **Gun Barrel Drills**.

Hollow Milling See **Milling** and **Trepanning**.

Hollow Spindle Lathes Lathes allowing tubes or shafts to be loaded through a hollow spindle at the back of the headstock so that these may be held by suitable jaws and taken through the headstock for machining. Thus the extremities of long work are machined without its being necessary to use a specially long lathe. The length at the back of the headstock may need to be supported.

Honing A form of grinding originally intended for internal work, but now used to give the final dimensions to a part and produce the required degree of mirror finish on the surface. It is not the same as internal surface grinding with an abrasive wheel, since in internal grinding the wheel has a small area of surface contact, whereas in honing the area is much larger. Honing is more economical than internal grinding because it involves less chatter, the time taken is shorter except when highly specialized equipment is needed, and the hone has flat contact rather than line contact. It can also be used to finish the external surfaces of cylindrical work at a fast rate.

The abrading tools or stones (sticks) are flat, rectangular and extremely short, up to 8 in. long or longer, while the width lies between $\frac{1}{4}$–$\frac{5}{8}$ in. They rotate at about 120 ft./min. with a longitudinal traversing motion in a widely-angled path, preferably 45 deg. The marks made on the surface are longer and more hatched than in normal internal grinding. The length of the hones is about half the internal

HONING

diameter of the work. A paraffin coolant is used, and the abrasive grit size is 36–180 for rough work and 300–600 for finish honing.

The stones clean themselves of swarf, 3, 18 or 24 being arranged in holders along the internal circumference of the work at equal distances apart. Three grinding motions are used, backward, forward and rotational. The holders are shoes with controlled radial movement from a common centre, and are fixed to the machine spindle, their longer axes parallel to it, their shorter axes radial.

The operation is mostly used for fine-finishing engine cylinder bores, diesel engine sleeves, small compressors for refrigeration, etc., ball bearings, taper bearings, and any bearings of fine finish and high quality. It is specially valuable for length of service and working efficiency. Work may be finished to ±0·0001 in. dia., precise circularity and straightness being given. Roughing brings the diameter to within 0·0005 in. or 0·001 in. of the final dimensions, and final honing is from 0·0001 in. on the smallest bores to 0·001 in. in bores of, say, 3 ft. 6 in.

Honing also rectifies dimensional errors, but not misalignment of a bore. Table IX gives the number of stones for each diameter. For maximum production the stones are cemented to sheet or hardened steel holders, but otherwise may be clamped or cemented into holders, or both. Hones may be made expanding to ensure maintenance of pressure on the surface. For steels, honing speed ranges from 70–150 ft./min. and 100 strokes/min. is average for a 5 in. stroke. The pressure ranges from 30–70 lb./sq. in. for roughing to 0–50 lb./sq. in. for finishing.

TABLE IX

Stones (Honing) per Dia.

Hole dia. (in.)	$\frac{1}{16}$	$\frac{1}{8}$	$\frac{1}{4}$	$\frac{1}{2}$	1	1½	2	3	4	6–8	12–24	36
No. of Stones	2	2	2	3	4	5	5	6	6–10	6–10	9–12	12–18

Honing machines are vertical or horizontal, and multiple machines carry as many as eight bores to be simultaneously honed.

Blind holes can be honed, and machines have been designed for taper bores. Diamond abrasive is extensively used for polishing very hard materials, but for intermediate operations a vitrified bonded diamond hand-hone is used, of 600 grit, measuring $\frac{1}{4}$ in. × $\frac{7}{16}$ in. × 4 in. See also **Electrochemical honing.**

Hook Angle See **Tool Angles**.

Hook Tool A type of tool employed in a vertical slotting machine and set transversely to the ram axis of the machine.

Horizontal Boring Machine See **Boring**.

Horizontal Lathe A type of boring machine with a vertical axis for machining the bores of cylinders of large size for engines, and also for boring rings liable to bend or warp under their own weight if placed on their sides in a lathe or boring machine of conventional type.

Horns In a planing machine, the levers of curved form pivoted at the side and tipped over by the tappets to provide the desired feeds to the planing tool, as well as the table-reverse.

Horseshoe Gauge A fixed gauge in the shape of a horseshoe used in measuring machined work. In effect it is a fixed caliper produced from a single piece of metal, the narrow ends measuring the part to be checked. Gauges of this type are used as standards in repetition work.

Hot Saws Circular saws designed for cutting metals hot in a suitable machine.

Hub A term sometimes used in the United States for **Hob** (q.v.).

Huntington Dresser A type of tool used for dressing grinding wheels, made up of a series of circular cutters carried in a suitable holder and located on a spindle. These cutters are of varying types and designs conformable to the particular purpose, and use discs of metal as the cutting-medium.

I

Increase Drill A form of twist drill (q.v.) whose increasing flue angle recedes from the drill point, the object being to clear the chips more readily.

Independent Chuck A chuck whose dogs or jaws move individually and without reference to each other, the opposite of a concentric chuck.

Index Dies In certain operations, an example of which is the nicking of discs of large diameter, use is made of a revolving index plate capable of taking the disc or other object past the punches.

These then nick the disc or produce a number of nicks every time the press makes a stroke.

Indexing A method of mechanically dividing the circumference of a circular or cylindrical component into a number of equal divisions, so that flutes, grooves, teeth, etc., can be machined, usually on a milling machine, with equal spacing between them. The mechanism used is termed the indexing or dividing head, and causes the work to rotate (if we consider as example a cutter with 20 teeth) 1/20th of the circumference after the completion of each tooth. A typical indexing head has a central main spindle carrying a worm wheel whose teeth engage with another worm directly below the spindle, and is carried by a shaft at right angles to the main spindle. At the other end of this second shaft is a spur gear whose teeth engage with those of a second spur gear mounted on a third shaft. This third shaft also carries a crank handle for rotating the mechanism when indexing.

The worm wheel has a specific number of teeth (perhaps 40), but only a single thread, so that the main spindle is given one complete revolution by turning the cranks as many times as there are teeth in the worm. Thus, if the gear has 40 teeth and it is desired to cut 10 flutes in a circumference, the number of worm wheel teeth is divided by 10 to give 4 worm turns for each flute. A single spindle revolution needs 40 worm turns. Attached to the main spindle is a circular plate containing circular concentric rows of holes, each hole the same distance from the others and the concentric rows all having a different total number of holes. There are various methods of indexing, namely plain, compound, differential and angular.

Indexing can be automatic and applied to almost any hand-operated turret lathe. It means longer time must be taken over setting, but gives higher production and minimizes the effort of the operator.

Modern dividing-heads do not embody mechanism for compound indexing, differential indexing being much simpler and more widely applicable. As an instance of this simplicity, compare the calculation for 57 divisions by differential indexing with that used in compound indexing.

57 divisions required. Round out to 60 for ease.

$\frac{40}{N} = \frac{40}{60} = \frac{2}{3}$. Index this 57 times.

$57 \times \frac{2}{3} = 38$ i.e. 2 short of the 40 needed for one revolution of the work. Thus, a change gear ratio of 2:1 is required to allow the index plate to catch up. One modern dividing-head has only a single row of 24 holes on the direct-indexing-plate on the spindle-nose, in place of the three rows formerly used.[1]

Indicating Gauges See **Gauges.** These show the departure from uniformity of size or form in a piece by movement of a pointer on a graduated scale or dial which indicates to the operator the amount of the departure or variation.

Inertia Disc Dampers In boring metallic components, a method of preventing vibration by inserting a batch of inertia discs of somewhat varying diameters into the hole of the boring tool. The vibration of the boring bar leads to the sliding of the discs into contact until they impinge upon the bar at irregular times, so minimizing the vibration.

In-feed In grinding parts of which some portion exceeds the diameter of the finish-ground work, a method similar to plunge-cutting or form-grinding (q.v.) on a grinding machine of centre type. The method is also adopted when it is necessary to grind a number of diameters at one and the same time, or when taper, spherical or other contours of unusual form have to be finished. The sole restriction on the sections to be ground in a single operation is the wheel width. There being no axial motion in relation to the workpiece, the controlling wheel is so placed that its axis is roughly parallel to the wheel axis, a small angle holding the part firmly against an end stop.

In-line The term used for work in the same centre or plane. To align different points, a known level surface may be measured; a fine chalk line made and tested with a spirit level; or in other fields, a theodolite is used.

Insert Countersinks Countersinking tools secured to a twist drill body and held fast at the required depth. The drill body pilots the tool and renders it rigid. Close to the bottom of the spindle movement the insert cutting-edge enters the work surface and produces the required countersunk hole. (See **Countersinking.**)

Inserted Blade Reamers See **Reamers.**

Inserted Chaser Taps See **Taps.**

Inserted-tooth Saws Circular saws in which the teeth are individually bolted to the plate by suitable means. They render it unnecessary to replace or re-tooth the entire saw when one or two of the teeth are fractured in service. (See also **Segmental Saws.**) Some **face-milling cutters** (q.v.) are also provided with inserted teeth.

Inserts Pieces of tungsten carbide or alternative cutting material mechanically held, brazed, soldered or welded into position on dies

---

[1] For this example the writer is indebted to Mr. James E. Shaw, Grad. Inst. Prod. E., of Shipley, Yorks.

INDEXABLE INSERTS

or cutting tools, and discarded when worn out, others being fitted in their place.

Indexable Inserts are rhombic in form, have an 80 deg. nose angle, and allow both turning and facing to square shoulders with the same tool. They are claimed to extend the versatility of throwaway tooling to all kinds of lathe work and to have a stronger toolpoint than normal copying or triangular inserts. They replace brazed tools and give all the advantages of indexable insert tooling. The holders, designed for simple insert indexing and unobstructed chip flow, are standardized in three shank sections, and the inserts themselves are available plain or with form-sintered chipbreakers on one or both sides.

Integral Pilot Counterbores Counterboring tools having two helical flutes designed to bore holes for small screws up to $\frac{3}{8}$ in. dia. They give precision of size combined with absence of chatter. (See **Counterboring**.)

Interchangeable Cutter Counterbores These tools comprise cutters of various diameters inserted interchangeably in a holder. They are driven by either a pin, a key or a spline, and are held in straight or tapered shank holders designed for severe service. They have a pilot (q.v.) of integral or interchangeable type.

Interlocking Gear Cutters Cutters of disc type revolving on axes at an angle to the face of the mounting apparatus, used for generating the teeth of straight bevel gears or pinions from a solid piece in a single operation. The cutting edges of the cutters have a concave surface to provide greater metal removal at the tooth tips. The work is secured in a spindle revolving in synchronous relation to the cutter-mounting device. The work-head and work take up position for roughing without generation, and machine to not quite the entire tooth depth. The work is then cut to the full depth and the sides of the teeth are generated. The work is disengaged, and both mounting apparatus and spindle descend once more to the roughing position, the work being simultaneously indexed.

Intermittent Feed As opposed to continuous feed in a machine tool, the provision of feed by a pawl and ratchet movement. The work can also be fed in this way.

Internal Grinding A method of finishing the surfaces of holes, such as the bores of ring gauges and jib bushes, the centre holes of gears, cylindrical cutters, etc., especially when the work has been previously heat-treated. The holes may be straight, tapered or both combined, while complicated hole shapes can also be ground. A special advantage of the operation is that parts may be precision finished from 0·00025–0·0001 in. In addition it is low in cost and provides a

INTERNAL HONING

satisfactory surface, and is often used to rectify small amounts of distortion in the internal holes of long thin work while providing a true surface.

A universal tool and cutter grinding machine with a special attachment is often used. In this an arbour or spindle driven from the normal wheel spindle is embodied, with the cylindrical grinding attachment to drive the work. Special horizontal or vertical grinding machines are also used. The work may be stationary, the wheel travelling anti-clockwise, or moved in planetary fashion against, and simultaneously traversed forward and backward over, the work. In other machines wheel and work rotate, the work also travelling to and from the wheel. It is preferable for the work to rotate, as the wheel is then firmly supported and does not experience the chatter caused by the rapid revolution of small wheels. Only for massive parts does the wheel reciprocate. The machines may be automatic or hand-worked. Planetary grinding machines are for massive or heavy work difficult to handle.

Parts that cannot be rotated are ground by vertical or horizontal spindle machines. Some awkwardly formed parts need an extensive swivelling motion and are ground in a machine that rotates the work, which is accommodated by an adjustable gap bed and is held in fixtures or other devices.

The wheel used should be softer than for cylindrical grinding as it is then less liable to spring, the wheel grade depending on the work and the machine's rigidity. High wheel speed is essential for satisfactory grinding, as it prevents rapid wheel wear, and harder wheels are used where vibration is greater than normal. A rigid machine gives greater accuracy and production. Machines for small work are ready-mounted. Wheel diameter is proportional to work diameter. See Table X. Speed affects bond, which must be harder as wheel speed declines. Holes not exceeding 2 in. dia. require a wheel $1\frac{1}{8}$ in. wide, but larger holes are ground in special machines with a wheel $1\frac{1}{2}$ in. wide. Traverse of wheel or work is governed by wheel width, being larger for roughing than for finishing. Coolant is as for cylindrical grinding. The maximum cut is not more than 0·003 in. deep, but coarse feeds in a modern, well-designed machine enable a good amount of metal to be removed.

Work can be internally ground in the lathe, a portable electrically-driven grinding machine being mounted on the lathe carriage. This is done only when the number of components to be internally ground is small. Chucking and centreless internal grinding machines are, however, mostly used.

Internal Honing See **Honing.**

Internal Threading See **Threading.**
Interrupted Cuts Cuts discontinuous in character, the tool contact being interrupted by gaps, which set up excessive chatter and wear in machining, and are often responsible for tool failure. In planing, for example, this can often bring the tool to a red heat at the nose.
Interrupted Thread In taps, the removal of alternate threads along the helix when the number of flutes is odd in order to mimimize friction and enable heavier cuts to be taken by each tooth. Interrupted threads are also used on occasion for comparatively large taps working on resistant metals.

J

Jaw Chuck See **Chucks.**
Jig Borer A machine tool for boring accurately up to millionths of an inch in guaranteed positioning.
Jig Bushes Standard bushes of three main types: (i) press-fit, drill-guiding, or boring bar, pressed directly into a jig; (ii) liner bushes or master bushes for guiding tools, and carrying replaceable bushes for this purpose—these are also pressed into the jig; (iii) bushes combining the two previous types and known as press-fit and liner bushes, which may be headless or supplied with heads, plain or knurled.

TABLE X

Wheel Diameter and Width for Internal Grinding

Work Dia. (in.)	Wheel Dia. (in.)	Wheel Width (in.)
Up to 1·25	Same as work	$\frac{3}{4}$
2	$1\frac{1}{2}$	,,
4	3	According to the work
6	4	,,
12	8–9	,,

Jobber's Reamer A type of chucking reamer with straight flutes having a length almost double the length of the drill itself.

Jobbers' Twist Drills Straight-shank short-series twist drills used in self-centring chucks, and normally having two helical flutes. They are less expensive and more handy for many purposes than Morse taper-shanked drills and are made in sizes up to $\frac{1}{2}$ in. dia. (See **Twist Drills**.) Being versatile, they are particularly applied to the general drilling of ferrous and non-ferrous metals and alloys. Their flutes are short and the relation of their diameter to length is valuable in ensuring that the drills are rigid in operation. They do not run ahead or snatch the material on which they work.

K

Kerf The cut produced by the teeth of a saw, whose width depends on the saw tooth-set.

Kerosene Cutting Fluids A type of paraffin used in machining as a coolant. See **Cutting Fluids**.

Key Chuck See **Chucks**. A type of jaw chuck (q.v.) whose square-head screws are turned by a square spanner or a key.

Keyway Milling Milling keyways such as those employed in arbours, which have considerable length. A form of milling cutter machining on the circumference gives both greater precision and output than ordinary end milling. See **Milling**.

Keyway Tool A tool for use in a slotting machine for the vertical machining-out of keyways, the wheel resting in a horizontal position on the machine-table. The tool may be of cranked form to enable the cutting-edge to stand clear of the shank, the edge having the same width as the keyway.

Knurling Increasing the hand grip of a tool or other circular component by providing it with milled ridges on the periphery to enable it to be rotated with greater ease. Sometimes, however, knurling is used merely to decorate a surface. Typical examples of knurled tools include gauges, thumbscrews, etc. Some knurling is done with dies of tool steel, and sometimes by other operations on one and the same piece.

L

Lands The raised parts between the flutes of a cutting tool such as a twist drill or reamer. The metal of the land is precision-ground to dimensions, and controls the final diameter of the hole. The entire land, has, however, to be relieved, as it is not wholly concentric and must therefore not be allowed to rub in the cut and lose temper through excessive frictional heat. Only the narrow edge or margin is consequently left concentric to ensure correct diameter. When a twist drill is having metal removed from its body to provide clearance, the narrow cutting land is allowed to remain to its original dimensions. If the drill lands are worn away, the only method of restoring their effectiveness is to remove the damaged portion.

In taps, reamers, milling cutters, broaches, etc., the lands are the top surfaces of the teeth and of full finished size, their width being calculated along the external periphery from the front teeth faces to the roots, the lands constituting the top or clearance surface behind the cutting edge, but not the sharper-angled teeth at the back of the tool forming a portion of the flute or chip clearance area. The usual bevel angle at the lead of a reaming tool is 45 deg., but this can be altered if the surface is desired to have a superior finish.

Lanolin A yellowy wool-fat, viscous and wax-like, readily emulsified with water and highly acid- and alkali-resistant, used as a buffer (q.v.) in honing to reduce or prevent vibration of the stones, and take up the shock and recoil of variations in power. It is expensive, but is nevertheless economical because of its long service. It flows readily under pressure, adheres to a metallic surface more effectively than mineral oils. It consists of cholesterol in various forms.

Lapping A polishing and truing operation using a tool known as a lap, comprising a metallic body of a soft metal such as lead, tin, brass, etc., as the support or matrix for the abrasive powder used. The lap has the same contour as the work in most instances, and produces work of maximum accuracy. It finishes plane surfaces or the internal holes or bores of cylinders, etc. The lap itself is a revolving, circular tool of metal, wood or other material, charged with an abrasive powder suspended in a coolant. Lapping gives a highly satisfactory surface, the work having a contour almost perfect so that the surfaces will mate together snugly and firmly if desired. It is now mostly a mechanical though originally a hand operation, and

LAPPING ABRASIVES

demands a high degree of skill. The lap material must be softer than the work, and the usual metals used are copper, brass, cast iron and low carbon steel. See Table XI. The softer the lap, the faster its cutting, the less its accuracy, the slower its wear and the brighter the finish produced. The greater the density of the material, the slower the cut.

Table XI
Some Materials for Laps

Material	Type of Operation
Brass	Ring gauges. Thin steel rods.
Copper	Finishing poor surfaces. Removing excess material. Thin rods.
Felt	Final surfacing of dies and moulds. Precision work on ball bearing raceways.
Fibre	Worms and worm gears.
Iron, Cast	Chasers, cutting tools, gears, gauges, plungers, crankshafts. Hand lapping. Maximum accuracy.
Lead	Badly surfaced ball bearing raceways. Large cylinders. Rough-finishing dies and moulds.
Leather	Ball bearing raceways.
Micarta (a synthetic insulating material with a mica base)	Worms and worm gears.
Steel, Low Carbon	Accurate finishing of small holes.

Lapping abrasives include diamond, garnet and emery dust, aluminium oxide and silicon carbide. Silicon carbide is commonly used for gears and hard steel components. Aluminium oxide is for lower carbon steels; and garnet or emery for reduction gears. Diamond powder is for exceptionally accurate small work, tools, etc. The range of abrasive available is from 60–1,000 grit, but in practice the coarsest grit used is 150. The finer the grit, the more accurate the soft lapping operations.

The abrasive particles are suspended in grease, oil, water, or a volatile spirit such as alcohol. The fluid chosen is termed the vehicle, and the heaviest body gives the best finish, but less material is removed per pass. Alcohol gives the maximum cut, but a less satisfactory finish.

LAPPING MACHINES

Charging the lap with abrasive for cylindrical laps on external work is done by a hard steel roller of diameter somewhat smaller than that of the lap. For internal work an arbour is passed through the lap. A rectangular piece of hard steel is coated with the vehicle, and the lap rolled upon this with sufficient pressure to load it thoroughly.

Lapping machines are divisible into those for lapping both flat and round work and centreless machines. The modern practice is to use a disc wheel with an extremely fine abrasive and a lubricant. The machines may be vertical or horizontal, the horizontal type being mainly used for the disc wheels. Centreless lapping is a variation of **centreless grinding** (q.v.), but gives diametral accuracy of 50 millionths in. and roundness within 25 millionths. It is used for high production of continuously-fed parts from $\frac{1}{4}$–6 in. dia. and up to 15 ft. long. For the long work a long bar feed is necessary. The operation takes off little material, so that work must be accurately round and straight before lapping begins. Typical parts centreless-lapped are pistons, piston pins, shafts and bearing races.

Other lapping operations include outer cylindrical surfaces, machine lapping between plates, centreless roll lapping, crankshaft lapping, piston ring outer-surface lapping, inner cylindrical- and flat-surface lapping, end-surface lapping, spherical-surface lapping, ball lapping, springlike-part lapping and gear lapping. Materials lapped include aluminium alloys, bronze, cast iron, powder metallurgy parts, stainless steels, tool and other steels.

Lard Oil A coolant used efficiently for only those operations where little heat is generated, in which instances it is entirely satisfactory. When pure it cannot be used for heavy tapping, drilling deep holes or threading difficult materials. In addition it is expensive but is usually diluted with a mineral oil such as paraffin, petroleum oil, etc., and used for automatic screw machines, drilling, reaming, gear-planing, etc., being thin enough to flow, yet viscous enough to film the machined work. Typical proportions are 30:70, 11:89, 25:75, $33\frac{1}{3}:66\frac{2}{3}$; lard oil:mineral oil.

Laser Beam Machining A means of removing metal from a workpiece by melting and vaporizing it under the influence of a narrow beam of monochromatic light of intense character. The name "laser" comes from the initial letters of the words "light amplification by stimulated emission of radiation". Its application is to materials that can be machined in no other manner to obtain the desired result. Particular uses include the boring of holes in extremely thin metal or material and machining with high accuracy and small cuts.

LASER BEAM MACHINING

The process necessitates a source of light, such as a xenon flash lamp or a linear arc discharge lamp and a "laser" of ruby or neodymium in glass. Lasers of neodymium are much superior to ruby and respond less quickly to variations of temperature. The operating temperature is usually that of a normal room. The laser is non-conductive electrically, its power being provided by a pulsed light flux with a d.c. power supply and a bank of capacitors, charged to 4,000 volts. The d.c. gives a pulsed light flux of 3,000 joules discharged in 1 millisec. through the xenon lamp. The 2–6 per cent neodymium-containing rod in glass is situated at the foci of an elliptical polished aluminium reflector, and almost all the lamp radiation is focussed on the rod, whose ends are reflectively coated. The light is emitted as a parallel beam, and a simple lens focusses it to give high power densities in small areas 1–6 in. from the lens. Additional equipment is required, such as a triocular microscope and a suitable workholder. Cooling by water or air to minimize generated heat is required.

Laser machining takes off only about 0·0004 cu. in./hr., much less than **electron machining** (q.v.). Parts can be machined to a dimensional accuracy of ±0·001 in., but despite its high cost, the process has advantages, among which are that it will machine virtually any material; there is no contact between tool and work; nor any large forces between beam and work; machining can be done in the air or in an inert gas, in a vacuum or through transparent fluids and solids. Exceptionally small holes can be produced and fine cuts taken. Materials of ceramic and other types readily impaired by thermal shock can be machined without difficulty.

On the other hand, the disadvantages, apart from high cost, include: limited applicability and small removal of material; the work having to be precision-aligned, which reduces the rate of production; the holes and cuts cannot be guaranteed uniform; the work possibly being injured by high temperature; and skilled operators being required. Characteristic machining work done by laser beam is drilling holes in ceramic materials and exceptionally hard metals such as tungsten. Holes have, for example, been drilled in zirconia that could have been drilled to the accuracy required by no other method without fracture.

Gas lasers using an infra-red gas beam focussed by mirrors rather than lenses is an alternative method, said to show great promise. They give an uninterrupted power output and a conversion efficiency between 13–20 per cent. They are low in cost and more convenient.

LATHE

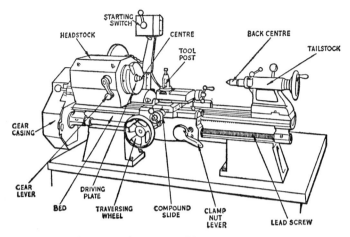

Fig. 19 The lathe and its component parts

Lathe A machine tool used to produce circular components. The workpiece is held between the centres of rapidly rotating and movable heads or secured to the **headstock** (q.v.) only by means of different forms of chuck. The cutting-tools are carried in a tool post in a slide rest, automatically or manually operated. The headstock and work-rests are carried by a bed or bearing surfaces.

Lathes are of various types. The **turret** or **capstan lathe** has a cylindrical head or turret containing a range of tools, any of which is put to work by revolving the turret. **Automatic** and **screw machine lathes** have their feed motions and operations automatically controlled. A lathe is known as a **turning lathe** when it machines the external surface of the work, and as a **boring machine** when it machines an internal surface. The bed is usually horizontal, the work rotating about a horizontal axis, but vertical spindle lathes and borers have the work fixed to a horizontal table rotating around a vertical axis.

The **lathe bed** is a horizontal raised platform with runways or grooves precision-machined to prevent the carriage or saddle carrying the cross-slide and toolholder from deviating from the parallel and so causing the cut depth to vary. The bed carries the working parts of the lathe.

The **headstock** (q.v.) is matched on the operator's right by a **tailstock** or loose headstock, whose centre is carried in an internally screwed barrel so that the tailstock cone can be adjusted to suit the tailstock setting. The cone-shaped centre of the tailstock carries the

LATHE TOOLS

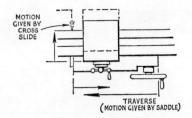

Fig. 20   Normal traverse motion of lathe

work, rotating it about a true axis. Many cylindrical parts turned in the lathe between centres have their ends drilled with central tapered holes to receive the centres, but in modern lathes some tailstocks have revolving centres. The tailstock is moved along the lathe bed longitudinally according to the work or to set it free when not in use.

The **cross-slide** travels horizontally along the bed to produce the tool feed, and transversely for the cut, the cutting-tool and its supports being attached to the slide. The feed mechanism controls the carriage traverse and sometimes the cross-slide cutting motion. Carried by the headstock is the main driving spindle whose power is supplied by electric motor or belt. Chucks for holding the work are located at the end of the spindle shaft fixed in the headstock, or alternatively a faceplate disc is fastened to the spindle, the work being clamped or otherwise secured to it by bolts passed through holes in the plate.

The carriage moves in either direction along the bed, inward and outward or forward and backward movement being given to the tool by the cross-slide moving at right angles to the bed length.

Some lathes have an additional slide to carry the cutting-tool and enable it to be swivelled (swung) horizontally and presented to the work-face at various angles. This is termed a **compound slide-rest**. The toolpost, usually cylindrical, is fastened to the slide. Various types of lathes are used for different types of work, and the materials to which they are applied include lead alloys, steels, plastic materials, etc., which can be bored, drilled, tapped, threaded, and turned. Lathes can also be used for grinding, honing, lapping, reaming, roller burnishing and other operations.

Lathe Tools Cutting-tools specifically designed for use in lathes, planers, shapers, etc., and used for both roughing and finishing, their cutting-edges being ground to the particular angles required by various materials. (See **Tools**.)

Lead Angle The helix angle of a screw thread measured from a plane perpendicular to the axis. It is governed by the lead of the thread or gear tooth when screw threads, worm wheels, etc., are machined. Applied to cutting-tools used for turning, the expression connotes the side cutting-edge angle. (See **Tool Angles**.)

Lead Screw The screw of a lathe designed for screw-cutting. It travels lengthwise in front of the lathe bed, and is sometimes called a **main screw**.

Left-hand Drills Twist drills, whose helical flutes run in the opposite or left-hand direction instead of to the right as in a normal drill, used in special machines, as when multiple-operation machines have the spindle revolving in the opposite direction to the usual, and in tapping after drilling or when drilling is combined with other operations.

Lifting Blocks Metal pieces inserted under the headstock or rest of a lathe as packing, so that the machine may be employed for a particular purpose, but not permanently. Sometimes these pieces are of wood.

Light Cut In machining metals, a cut that produces only small, thin chips, but as with **Heavy Cut**, the term is not an accurate one.

Light-wave Micrometer A measuring instrument combining the high sensitivity of light waves with the convenience of the micrometer (q.v.). The wave bands travel past a reference line when the work to be measured is undergoing pressure, and the wave micrometer measures the diameter, thickness and length of the work between flat, parallel surfaces, avoiding the production of the impressions left by many other measuring devices on the work-surfaces. The instrument is specially serviceable when extremely light and soft substances or parts have to be measured, since the pressure employed is slight. Its applications are, among others, to cylinders, screw-threads, rubber, leather, textile materials, paper, thin and delicate wires, dental fillings, hairsprings, etc.

Lip Relief Angles The lip relief angle of a twist drill depends on the diameter of the drill, the work-material and its condition. It may range from 7–20 deg. for hard or tough metal; 12–26 deg. for soft or easy machining metal; or 8–24 deg. for cast iron or annealed steel. See also **Tool Angles**.

Liquid Jet Finishing A novel means of finishing a workpiece by cutting or boring with a high pressure liquid jet, produced in a blast nozzle 2–3 mm. dia. bore, at pressures of 100–900 atm. The jet is directed on to the work and contains a particulate material mixed with the liquid. A German invention, it is covered by Br. Pat. 1,148,072.

Live Spindle A spindle, such as the mandrel of a lathe headstock, capable of communicating its own motion to some other body.

Loading The condition of a grinding wheel whose surface has become clogged with minute particles of the work material so that it cuts less freely. This fault is usually encountered when the wheel is too hard, improperly bonded or badly dressed, excessively stressed, ill-fitting, or badly chosen.

Low Helix Drills See **Slow Helix Drills.**

Lubricants See **Cutting Fluids.**

M

Machine Countersinks Countersinking tools having radial relief and four flutes, with centring angles of 60 deg., or, for flat-headed screws, 82 deg. Bevelling or removing rough edges calls for an angle of 90 deg., but it is possible to modify these angles as required. See **Countersinking.**

Machine Lapping between Plates See **Lapping.** A method of finishing external cylindrical surfaces where large quantities are required and the form is suitable, in which instances it is economical. It gives plug gauges, piston pins, hypodermic plungers, ceramic pins, and many other cylindrical parts a finish of from some thousandths of an inch up to 3–4 in. dia. and from $\frac{1}{4}$–9 in. long. It can be used for either soft or hard materials.

Main Screw See **Lead Screw.**

Major Diameter The largest diameter of a screw thread, sometimes termed the **outside diameter.**

Mandrel (i) A shaft, precision-turned, on to which parts previously bored are mounted; (ii) the spindle of a lathe headstock (q.v.); (iii) a rod maintaining the hollow or cavity in metals subjected to heavy rolling pressures; (iv) a partly-split shaft that can be expanded by a plug of taper form driven into it. Mandrels used in machining usually have their ends centred. Those used in the cylindrical grinding of rings are of two types, **tap-up** for use where the inner diameters have been left unground or are not precision finished; and **pilot** for use where the inner surfaces have been precision-ground, the mandrels being placed on these inner diameters so that the rings may be externally ground. Expanding mandrels are sometimes applied to improving the concentricity of a wall thickness when this could not be achieved with the normal solid mandrel.

Manganese Steel, Drilling Special drills are designed for drilling Hadfield austenitic manganese steel (12–14 per cent manganese). The drills are short and stubby, having a large cross-section and a specially thick web. They cut without a lubricant and require heavy machine-power to drive them. They will also cut other material liable to work-harden, and difficult materials such as white iron.

Maskants In **chemical machining** (q.v.), substances designed mainly to withstand the etching action of the fluid used. They may be of rubber base or resinous. (See **Chemical Machining** and **Chemical Contour Machining**.)

Master The tool used in **chemical blanking** by the photo-resist process. (See **Chemical Machining**.)

Master Gauge A gauge used to test other gauges for accuracy and for no other purpose. Virtually no allowances for wear are made in these master gauges.

Master Plate A plate of steel enabling holes to be precision-bored in jibs, fixtures, etc. The plate is given as many holes as the work, then mounted on the machine face-plate, and after a particular hole has been bored, both plate and work are moved and relocated for the boring of the next hole. The method has now been largely replaced by the jig borer (q.v.).

Matched-piece Lapping In **lapping** (q.v.), two parts to be lapped and separated by merely a single film of abrasive material are rubbed together so that each drives the particles of abrasive onto the opposing surface. This enables those unevennesses that stop the surfaces from accurately mating to be eliminated. In this way tight seals for gases or fluids can be lapped so that gaskets are no longer needed, while in some instances piston rings need not be employed in cylinders. Another application of the process is to mating a pair of gears or to fitting parts having tapers, as in valves.

Mentorial Electronics A term coined to express a method of control in the maximum-production machining of intricate components for the automobile industry. Each station of a transfer line is linked to a machine controller, which in turn is linked up with a small, general purpose computer to monitor and report on the performance of the machine.

Metal Bond Wheels Grinding wheels used in **electrochemical grinding** (q.v.). They have a layer of metal taken off to expose the abrasive grains on the wheel surface, from which the metal is anodically dissolved by reversing the electric current.

Metallurgical Burn See **Grinding Burn.**

Microdrilling The drilling of small holes (0·001–0·375 in.) by special machines, tools, methods and skills. The holes are usually drilled by machines mounted on a bench, but larger machines may employ chucks or collets to hold the drills. Twist, spade or special drills may be required, and exceptional care is taken to grind, centre and feed them owing to the restrictions in chip removal space and the heavier pressure on the ends of the drills, as well as the greater longitudinal and torsional deflections. (See **Drilling.**)

Micrometer Gauge The micrometer gauge is made up of a frame carrying a cylindrical or other form of anvil or fixed jaw, working in a fixed locknut and used for measurement. A suitably-threaded spindle works in the locknut, and both locknut and spindle are secured to and contained by a sleeve or barrel. The sleeve itself is enclosed in a knurled thimble with a chamfered edge, and when the knurled portion is rotated by thumb and finger it rotates the threaded spindle, the frame being held stationary for this purpose. (See Fig. 21.)

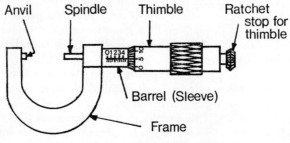

Fig. 21   Micrometer gauge

A single turn of the thimble advances the spindle towards the anvil, the lines and numbers on the sleeve indicating how many turns have been made. Each fourth graduation is successively numbered from 1 upwards, and since each mark represents an advance of 0·025 in., the graduation marked "1" signifies 0·025 × 4 = 0·100 in., "3" signifies 0·025 × 4 × 3 = 0·300 in., and so on.

The chamfered edge of the thimble is divided into 25 parts numbered at every fifth mark: 0, 5, 10, 15, 20, and back to 0. Each such division is equivalent to 0·001 in. The precise width measured is

MILLING

found by addition of the reading on the barrel, to the nearest 0·025 below, and the reading on the thimble edge.

Some micrometers have a vernier scale so that finer measurements can be made.

Milling A method of forming steel or other materials by presenting the work to be machined against a revolving cutter, i.e. a circular disc whose periphery is provided with teeth or cutting edges to which a specific form has previously been given to suit the form of the work it is desired to produce. These discs or **milling cutters** are usually of high speed steel, hardened and ground to suit the material to be milled, but many cutters are of composite type, a carbon steel body holding renewable teeth, sometimes of tungsten carbide or other cutting alloys. They do not run at a specially high speed, but produce more quickly and accurately than either planing or shaping the precise surface contour desired. (See **Milling Cutters.**)

Flat surfaces are usually milled with **face cutters** whose teeth are on the cutter face, or with **circular cutters** whose teeth lie on the periphery. These may be parallel to the cutter axis or at an angle to it, or the cutting edges may lie in helices like large pitch threads on the surface. Rectangular gashes or grooves can be milled by cutters with peripheral teeth and teeth on both faces. For gashing or grooving, special cutters are designed with a particular contour. Milling is extremely important and economical for a large number of operations, while special machines are used for many operations.

The **milling machine** rotates the tool and feeds in the work, which is secured to a movable table and automatically fed in. The milling cutter is mounted on a horizontal or vertical arbour on the machine spindle, and as the spindle revolves, the teeth in succession make a cut, the work moving in at each tooth-incision. Work-feed and speed of cutter are dependent on cut depth and material cut, as well as the dimensions and size of the cutter. Consequently the machine includes gears in the body to modify both feed and speed as required.

The machine gives the components accuracy of form, and is applied to milling flat surfaces, keyways, special cutters, the flutes of twist drills, the teeth of metal-cutting circular saws, the grooving of taps and the teeth of gear wheels. It feeds longitudinally, vertically or across the width. The knee or support for the work-table can be elevated or depressed in relation to the work height. Drive in most modern machines is by electric motor.

The two principal types of milling machine are the *knee type*, such

MILLING

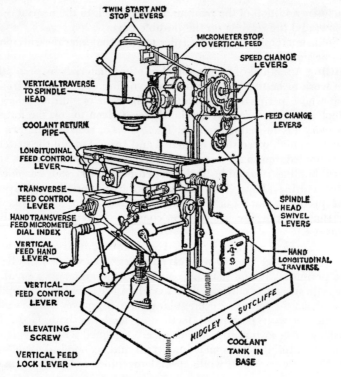

Fig. 22   Milling machine of vertical arbor type

as the *knee and column* machines, and the *bed type* which has no knee, the cutter moving down towards the work. In both, however, the work-table moves longitudinally in to the cutter. There are also special machines constructed for particular operations. The knee machines have a base; a column containing the spindle and its driving mechanism; an overarm giving support to the cutters on their arbour; a knee secured to the column, on which it moves vertically; a saddle moving horizontally on and being supported by the knee; and a work-table carrying the work and moving horizontally at an angle of 90 deg. to the saddle motion. Machines of this type may be horizontal or vertical. The first horizontal type having power feed to the longitudinal table movement only is known as the *manufacturing type*. The second is the *plain milling machine*, with power-feed longitudinal, transverse and vertical. The third is the *universal* with power-feed like the plain miller, but with also a swivel-

MILLING CUTTERS

ling table enabling inclined surfaces to be milled. It has usually a dividing head, for **indexing** (q.v.).

The vertical knee machines have only power-longitudinal feed to the work-table, combined with the transverse and vertical table movements.

See also **Plano-miller.**

Knee machines lack rigidity, and milling therefore produces a degree of deflection, minimized in some instances by the interposition of a support between overarm and knee. They have a limited range of speed, feed and cut depth, and if these limits are exceeded the cutter may chatter and impair the work and tool service-life. On the other hand, the bed machine type is of higher first cost. Automatic controls may be mechanical-electric, mechanical-hydraulic, mechanical-electric-hydraulic, and numerical.

Milling Cutters Tools used in **milling** and **reaming** (q.v.) and of various types. The ordinary **plain milling cutter** is cylindrical

Fig. 23   Plain milling cutter

in form with teeth on its periphery only. These teeth have equal distances between them, and their edges may be straight and parallel to the cutter axis, straight or inclined to the axis, or curved and helical in form. The straight type are little used today. This cutter

Fig. 24   Side cutter

127

MILLING CUTTERS

roughs or finishes a piece by removing metal from its surface, for which purpose it rotates. Helical teeth cut more smoothly, especially on wide flat surfaces, and give a better finish. They may be used to cut either left- or right-hand, and if more than 4 in. wide, are often built up in interlocking sections, while those more than 5 in. dia. have inserted cutting blades.

Fig. 25   Face cutter

Side and **face cutters** have teeth on one end only, but some have them on the circumference also, where they are equal to or less than half the cutter diameter. Teeth may be parallel to the axis on the

Fig. 26   Slotting cutter

circumference and radial on the face. They cannot be reversed and must therefore be either left-hand or right-hand. **Slotting cutters** have circumferential teeth only, but are much narrower than the plain milling cutter, the teeth up to $\frac{3}{4}$ in. wide being straight, and helical for widths above that, while their sides are slightly concave to give clearance. They are used for cutting slots with great accuracy.

Straddle mills have teeth on the face or circumference and on both sides also, being used in pairs, with packing pieces between them, to mill surfaces a specific distance apart. Normally they have helical teeth for widths above ¾ in. and straight teeth for smaller widths. They do not cut slots so accurately as slotting cutters, and have to be resharpened on sides as well as circumference, so that the tools become thinner with each resharpening, but they are often used in combination with other types of cutter for milling special contours in a single operation.

Concave and **convex cutters** are used to produce concave slots or convex edges, and are given machine relief to enable them to be resharpened by grinding without great modification of the form they produce. **Involute gear cutters** have the shape of the spaces between gear wheel teeth and so produce teeth of their own form

Fig. 27   Concave cutter

in gear wheels. Eight cutters comprise a set, and these will cut from 12 teeth upwards, or a **rack** (q.v.) if desired. Sets of 15 will cut gears closer to the theoretical involute form.

Single angle cutters have teeth on the side, the sharp corners being usually rounded off to lengthen the service-life of the tool. They are made to any angle and either left- or right-hand. **Double angle cutters** have the two faces unequally angled, and are used in cutting helical teeth, being either left- or right-hand. **Equal angle cutters** are similar in design to the above, but the nominal angle of the cutter is between the two conical sides. They are made to any angle, usually 45, 60 and 90 deg. and have no hand.

End mills have teeth on both face and circumference, and are used for facing cuts or circumferential cuts or both. They are made in sizes of $\frac{3}{16}$–4 in., but larger ones can be obtained. Those above 2 in. dia. usually have inserted-teeth of tungsten carbide or high quality high speed steel. They are as short as possible, left- or right-hand, with helical teeth, and are applied to surfaces not easily milled

MILLING CUTTERS

by cutters carried on an arbour. Some mills have two flutes for milling out a groove with shallow cuts, and need no preliminary drill incision. They range from $\frac{1}{4}$–$1\frac{1}{2}$ in. Relatively deep cuts are taken if

Fig. 28   End mill

the mill has a multifluted end. The shanks may be either straight or Morse taper, the former being the more widely used.

Shell end mills have circumferential teeth and teeth on one end, and are designed for arbour mounting. They are simply a larger end

Fig. 29   Shell end mill

mill than the conventional, but have cutting flutes longer than half the cutting diameter, being made to cut either left- or right-hand and have no shanks. Most have helical teeth. Minimum dia. is usually 1–$1\frac{1}{4}$ in., and maximum more than 6 in. Formed cutters or relieved cutters mill curved surfaces and irregular contours, many being of

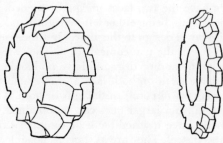

Figs. 30 and 31   Formed cutters

intricate and special form. They have a radial profile, and special methods are necessary for their relief.
Mineral Oils See **Cutting Fluids.**
Modified Underpass Shaving See **Shaving.**
Module A unit or standard of measurement.
Morse Taper Shanks Standard measurements used for the taper shanks of twist drills (q.v.), each taper being designated by a number (0–7). They govern the amount of taper per inch, and the dimensions of shank, tongue, keyway, etc. They are also used for the shanks of reamers (q.v.) and make it possible to interchange drills, etc., on the various machines.
Multiple Operation Machining The machining of materials by secondary operations, carried out *after* the original series of operations, to provide an absolutely efficient machining cycle, so greatly economizing in time and labour. The number of operations performed is virtually unlimited. Standard operations are mostly done with basic equipment, but a wide range of special fixtures, attachments and changes in design cover virtually all machining work. The machines included are automatic turret lathes, multiple-spindle chucking machines such as screw machines, single-spindle bar machines, multiple-spindle bar machines, manual turret lathes, etc. Many jobs call for as many as ten different operations on one workpiece, some of which are performed more than once. The technique is, however, rarely used for parts capable of being machined with fewer than three tools or for fewer than 10–15 parts of identical design and character.
Multiple Thread Cutters These have ring-shaped rows of teeth correct in contour and pitch, but without a lead, and are used for milling threads (see **Threading**). They cut on only the sides and root of the thread, having 2–3 pitches longer than the thread to be formed, or alternatively the work is rotated more than 1·1 turns/min.
Multi-spindle Honing Machine A British machine specifically designed for inclusion in transfer lines to hone the cylinder bores in engine crankcases and similar components produced in quantity. It hones bores up to 5 in. dia. × 12 in. long. Each spindle has its own power-source, so that the spindle stops immediately the bore has been honed to the right size. Adjacent bores in a cylinder block may be simultaneously honed.

N

Negative Rake See **Tool Angles**.
Neodymium-in-glass Lasers See **Laser Beam Machining**.
Nominal Size The term used for subdivisions of a unit of length not specifically limited in accuracy but approximating closely to a standard dimension.
Nose Radius See **Tool Angles**.
No-wear Electrical Discharge Machining A method of roughing a work-piece by **electrical discharge machining** (q.v.), using electrodes of graphite at positive polarity, high current density and low pulse frequency. It is not employed for finishing operations, but as there is no wear on the tools the process is often used for taking off considerable amounts of metal. A proportion of this metal having been liquefied or vaporized, it solidifies on the electrode in a thin, adherent film, which is not allowed to thicken sufficiently to change the machining circumstances too greatly. The process does not give low-micro-inch surfaces, and is much lower in cost for roughing die cavities than ordinary die-sinking.
Numerical Controls Machine tools in the modern world are being more and more controlled by electronic remote control. By this means the tool-motions when cutting, as well as those of dividing heads, cross-slides and speed and feed, are previously programmed and punched out on a tape, which is passed through an electronic device and ensures electronic control of every machine motion. As a result the number of tools stocked is radically reduced, since no templates are required, while the operation is so closely regulated that even to alter speed and feed necessitates the production of a new punched tape.
Nut Tap A type of **tap** (q.v.) specifically designed for cutting the nuts of vertical-spindle machine tools. Normally they are manufactured with a taper in the thread angle enabling the tool to cut a full form thread. These taps are longer and have longer thread and shank lengths than standard taps for hand use. They are now obsolescent.
Nylon Wheel A type of grinding wheel, impregnated with an abrasive material and composed of nylon, used for grinding aluminium sheets to give them a high polish, but not for roughing.

O

Off-centre Grinding Machines Machines used for grinding internal diameters. The rolls are so located that the centre lines of the component being ground and the wheel are above that of the controlling roll. This system of spatial relations between them ensures that each succeeding component occupies the same centre line irrespective of external diameter variations. Consequently close limits of size can be achieved, while numbers of parts can be ground simultaneously to the same dimension. The machines are primarily designed for centreless grinding.
Offset Boring Heads See **Boring.**
Oils, Cutting See **Cutting Fluids.**
Oil-tube Reamers The same principle as above is applied to **reamers** (q.v.) and can also be applied to both solid and inserted-blade types, the solid up to $1\frac{1}{2}$ in. dia., the inserted-blade above this. They are mostly used for horizontal work and provide a satisfactory finish.
Oil-tube Twist Drills Drills having copper tubes inserted in the drill body or holes centrally drilled in the solid body, the better and more usual method. The drill is then twisted to give the desired helix angle to the flutes. The holes are connected up with the drill point so that a coolant may be supplied to the lips and thence flow back along the flutes, thereby washing away the chips and extracting heat from the cutting point. They are sometimes used for deephole drilling. The drills are used in turret lathes or multiple-operation machines, or with power-drive in special sockets. They may be either straight- or Morse-taper-shanked, and are applicable to both horizontal and vertical drilling or wherever the provision of coolant to the drill lips proves difficult. They are advantageous when cutting hard materials, and allow of a higher rate of feed. (See **Twist Drills.**) Taper-shank are better than straight-shank oil-tube drills as they give better concentricity, higher speeds and feeds and greater drill rigidity.
On-centre Centreless Grinding Machines Machines used for the centreless grinding of internal diameters. The axes of work, abrasive wheel and controlling wheel lie in the same horizontal plane and full support is thus given to the work. They are applied to work having thin walls where precision grinding and minimum distortion are required.

One-plate Lapping Machines Machines designed for lapping flat surfaces in which only one cast iron lap is used with uncompacted abrasives. The lap rotates in a similar manner to a grinding wheel, and has a pressure of 1–10 lb./sq. in. or above when exceptionally light components are required. Copper laps are also used or those of bonded abrasive, diamond abrasive being popular for the softer metals. See **Lapping**.

Open-side Planing Machines Planing machines having a single vertical pillar and restricted to three tool-heads, two on the cross-rail and one on the pillar. They are lacking in rigidity as compared to double-pillar planing machines, but can be had in a considerable number of sizes. If necessary special holders can be used to take additional planing tools. (See **Planing**.)

Optical Drill A British type of optical drill with ×10 magnification developed for printed-circuit applications. Accurate drilling of holes free from burrs, with good surface finish, and working directly from a negative directly on to glass-based circuit boards, gives through-hole connection without fatigue in continuous production. The drill also produces prototype boards and templates for conventional production drilling. The optical system is an inverted telephoto arrangement focussed by a milled ring. Spindle speed is 20,000 r.p.m. and chucks or collets are accepted with capacities of 0·5 mm and 5 mm. Automatic feed.

Optical Flat A necessary tool in the machine shop, used in maintaining the accuracy of such measuring tools as gauges, micrometers, etc., etc. It immediately indicates worn areas in units of 0·00001 in. and even finer, yet is exceptionally simple and direct in use. It is in effect a piece or plate of hard, transparent glass or quartz so precisely ground and polished that one of its surfaces is, to all intents and purposes, absolutely flat, and the accurate flatness of this when laid on a surface indicates departure from flatness in the surface tested. Light waves of monochromatic type are used in a special neon- or gas-filled tube. When the optical flat is applied to a surface almost flat, or polished and reflecting, it shows at once bands of alternate dark and light tints. If these are straight, they indicate that the work-surface is of the degree of flatness required, whereas if curved, they show that flatness is not sufficient. The bands arise from the interference of the light reflected from the gauge surface with that reflected from the undersurface of the flat. The number of bands indicate the unit difference in thickness between the standard represented by the gauge and the workpiece.

Outside Diameter See **Major Diameter**.

P

Pack Head Pilot A type of **pilot** (q.v.) applied to a boring bar, using the work as a support. Inserts of nylon or bronze to the number of four provide the bearing surfaces. The pilot is a single cylinder extending beyond the cutting end of the bar to give the tool edges support and so eliminate sideways deflection.

Packing Piece A thin piece of flat wood or metal inserted between metal parts to bring them up to a required dimension or to provide a base for support in planing machines, etc.

Pad Chucks See **Chucks.** A type of chuck having a square hole of the same dimension as the tool.

Pad Saws See **Saws.** A type of saw with a narrow tapering band sliding between a hollow handle and held in place by set screws. It is small in size, and used for sawing curves of small radius or holes in the centre of a component.

Pantograph Milling Milling a component to a specific form from an enlarged master template of the same form. The operator guides a tracing stylus about the template (q.v.), the embodied cutting tool repeating the form of the template on a smaller scale. The template is often about 3–5 times larger than the work, enabling any mistakes in it to be proportionately smaller in the finished part. The reduction is achieved by a machine known as a *pantograph*, a mechanism based on the geometry of a parallelogram, compelling a point to repeat on any desired scale the path traced by another point. (See **Profile Milling.**)

Paraffin Cutting Fluids (See **Cutting Fluids**). Fluids having a base of paraffin mixed with oil or in the form of an emulsion, used in machining as coolants. For example, a typical formula is paraffin 20 gall., oleic acid 4 gall., denatured alcohol 2 gall., caustic soda 1 gall., plus water in the ratio 24:1, water:oil. This is emulsified and often phenol disinfectant 1 gall./1,000 gall. of emulsion is added. In deep hole drilling a paraffin-base oil of 100–125 viscosity at 38 deg. C. (100 deg. F.) is sometimes employed for aluminium. (See also **Kerosene.**)

Parting-off Tools Tools designed to part off sections of bars or ingots or to produce narrow recesses of keyways, and having therefore to be prevented from rubbing against the walls of the cut. They have a side clearance angle similar to the front clearance angle (see

Tool Angles) and a small or even 0 deg. rake to prevent the tool from digging into the work. The tools are often slightly rounded at the corners to give better chip clearance for machining narrow grooves. If the grooves are to be cut in face-plates or chucks, one side has a larger clearance angle than the other. For cutting difficult materials they are often made of cobalt-tungsten high speed steel, sometimes in the form of tips butt-welded to carbon steel shanks.

Peening Mechanically hammering with the narrower end or "pane" of a hammer or roller to smooth the surfaces of metals.

Percussion Drill A type of twist drill (q.v.) specially developed for drilling stone, brick, concrete and other hard materials. It has a special point obtained by grinding on the flat side of a grinding wheel, and can be used in a compressed air hammer, an electric drill, or as a simple hand tool.

Peripheral Milling The generation in a milling machine of machined surfaces by a **milling cutter** (q.v.) in which the teeth lie on the periphery. The cutter axis is parallel to the work. Great accuracy is obtained with this method of milling, especially in the cutting of keyways, while long keyways are also cut at a faster rate than by using end-mills. On the other hand, face-milling is better than peripheral milling for machining ordinary flat surfaces in that more material is taken off per min., while recesses of complex form and pockets are better machined by end-mills (q.v.). (See **Milling**.) The method is also used when economical for producing formed surfaces and those surfaces with an intricate shape or more than one angle, and is specially suitable for machining deep slots where rigidity is required. Some cutters for peripheral milling have teeth on both periphery and sides for special applications, especially the cutting of slots.

Care is taken to avoid faults, such as excessively high or low cutter teeth, distorted or chattering arbours caused by inadequate diameter, and unequal distance between the cutter teeth, resulting in bad work surfaces. Indeed, all these troubles impair the finish of the surface milled.

Pillar Drill A type of drilling machine in which a central pillar or column is carried by a suitable base.

Pilot A cylindrical device that guides a cutting tool into a hole to be bored or drilled, so giving it rigidity and ensuring that it does not move sideways owing to deflection. It is used for drills, boring bars, reamers, etc., in various designs. The pilot is normally manufactured from hardened and ground steel and must be resistant to wear, for which reason it is sometimes plated on the surface with chromium, or wear-strips of alloy type to the number of 5 or 6 are inserted around the tool. These wear-strips may be cast, wrought or hardfaced.

Broaches usually have a front and rear pilot portion with cutting-teeth respectively smaller and larger than the rest when the broach is pulled through the work. In countersinking, tools of insert type employ the drill body as a pilot. Many counterboring tools have pilots for stabilization of the cutting-action and to reduce chatter. These may be one with the cutter or independent and interchangeable, chosen to suit the operation. The pilot must not be too small for the tool, its diameter being preferably equal in diameter to the root diameter to ensure that chips do not clog in the space between cutting tool and pilot, as this may lead to tool fracture. The clearance between pilot and hole should be adequate to enable the pilot to revolve without sticking, but not so great as to set up vibration. 0·001 in. on diameter up to 0·007 in. should suffice for pilots of 0·125–3 in. dia. Pilots for counterboring should be employed for only holes not below 0·25 in. dia. Some integral pilot counterboring tools have a pair of helical flutes for holes to take screws up to 0·375 in. dia. where accuracy and lack of chatter are essential.

Pilots are also used in grinding rings by cylindrical grinding, in which operation pilot mandrels are used where the inside diameters have to be externally ground after precision finishing.

Piloted **ground thread taps** (q.v.) are also applied when it is essential for threaded holes to be concentric. Such taps are specially ordered to suit the user's needs.

Pipe Reamers Reamers (q.v.) of short and stubby type normally applied to ream pipe fittings with a tapered hole so that they may be finished by a tap. The reamers range from $\frac{1}{8}$–2 in. dia., and their amount of taper is $\frac{3}{4}$ in./ft.

Pipe Threading The machining with taps of threads in tapering pipes. (See **Threading**.) The machines employed are akin to those used in general tapping, but greater power is required and the number of teeth cutting simultaneously is greater, so that positive lead control is required. Solid adjustable or collapsible taps are used.

Pivot Drills See **Spade Drills**.

Plain Face Abrasive Wheels Contact wheels used in abrasive belt grinding (q.v.) and usually made of rubber, for grinding and polishing flat surfaces. They allow controlled penetration of the abrasive grit, and have a hardness of 20–40 durometer.

Planing A machining operation using single-edge cutting tools to smooth a rough, flat surface or for other work. The tools are clamped into position in a toolbox or toolhead, or rigidly held by other means. The machine has a bed, along which a moving carriage or table

PLANING

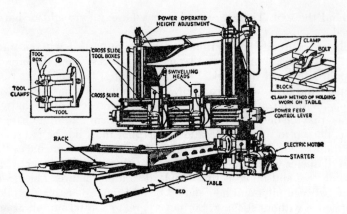

Fig. 32 Planing machine

runs on flat surfaces or vee-shaped grooves. The work is held securely on the table by clamping devices or chucks, and the table reverses. The stroke length can be adjusted by steps or dogs at the sides of the table.

Two vertical housings or standards like rectangular columns or pillars have their upper ends connected by a cross-beam, and are secured to the machine bed. These form an arch beneath which is a cross-rail or slide attached to the pillars, carrying one or more toolboxes and their tools. This rail can be elevated or lowered vertically on the pillars either manually or automatically. Toolboxes and tools may also be mounted on the pillars. Those on the cross-rail are moved vertically or horizontally by a slide. The feed is automatic. Toolboxes fitted to the pillars are used when vertical surfaces have to be machined, and are usually fed automatically.

The work travels in a straight line, the tool for horizontal surfaces being presented vertically to the work, then kept rigidly in position, moving laterally at the end of the stroke by a distance equivalent to the feed, so that as the tool moves laterally, the surface is levelled.

Planers are always used on large work where the plane surfaces to be machined run longitudinally, but are sometimes used for small work by stringing the parts out in a row end to end along the worktable length, so that each is machined in succession as the table travels forward. Having finished its forward stroke, the work-table now reverses to save time, the return non-cutting motion being quicker than the forward. However, in some modern planers machining is done on both forward and reverse strokes by double-cutting tools in specially designed holders carried in a special head, in which case the reverse speed is no greater than that of the forward stroke.

In ordinary planing the portion of the toolbox holding the tool rests firmly against the toolbox while cutting, but pivots on a horizontal pin on the return stroke so that the fixed tool shall not trail over and injure the work.

Two tool-settings are required, one to control the depth of cut, the other to control the feed. In roughing-cuts steel is removed as quickly as possible, and preferably at one stroke. The inevitable rough surface produced is then finished with a different cut depth and form of tool.

The machine is mostly electrically-driven, but hydraulic drives are also used. Open-sided planers have only a single column or pillar, the cross-rail extending over the table on the cantilever principle. These machines will accept wider pieces. In modern planers of either type lubrication is effected by a pressure system, oil being forced to the required areas. Speed-changes are usually provided by suitable driving mechanisms, countershafts or variable-speed electric motors. Solid stops in the table withstand the lateral thrust of the tool, and so minimize spring when heavy cuts are taken.

Many types of fixtures clamp down the work, but the common practice in many machine shops is to secure the work to the table by bolts, shims, packing pieces, etc. Multiple-planing is done by three or four tools planing simultaneously. Finish-planing must be specially accurate, and to correct elastic distortions produced by heavy clamping during the roughing-cut, the clamp pressure is often slightly relaxed. Long work of large and weighty type may bend to some extent under its own weight, and special support must therefore be provided at suitable points by small jacks or hardwood blocks.

To relieve stresses set up in planing cast parts, the work is sometimes allowed to stand after rough-planing and before finish-planing to allow the stresses to even themselves out naturally. If delivery times allow, any period from two to three weeks may be required for this.

Planing Generation A method of cutting gears normally limited to those about 35 in. dia. or above, or to diametral pitch coarser than $1\frac{1}{2}$. The cutting tools used have straight edges and are carried by a reciprocating slide supported by the cradle face and united to a connecting rod. (The contours of the teeth are formed by rolling the work with the generating gear.) Both straight and curved tooth bevel gears can be cut. A number of passes are needed to finish the gear, this number depending on form and depth of teeth. The tools are low in price and uncomplicated.

Plano-miller A **milling machine** (q.v.) similar in principle and functions to a planing machine, but machining work that could otherwise only be carried out in a planer. Its cross-rail can be elevated or depressed and carries the tools and their holders as well as the saddles, all being supported by strong and rigid vertical pillars or standards. There may be one cutter head or more on the cross-rail and two on the pillars. Each cutter head is independently driven, and the work-table does not reciprocate, but feeds forward gradually.

The machine is sometimes termed a **slab miller,** and is used for heavy work, especially when milling cutters have to machine plane surfaces. The tools usually have inserted teeth of cobalt high speed steel or tungsten carbide tips.

Plasma Arc Machining A form of machining in which a jet of high-temperature ionized gas is employed to liquefy and thrust out the metal before it as it proceeds along a prescribed course. The advantage of the process is that it is applicable to virtually all metals, and is specially suitable for producing profiles in aluminium alloys and rust- and heat-resisting steels. The work is brought to a high temperature by a narrow stream of electrons and by the energy transferred from the gas. Carbon and alloy steels are normally machined or flame-cut with a gas of mixed nitrogen and hydrogen or compressed air. The more difficult metals are machined with mixtures of hydrogen with either nitrogen or argon flowing at 70–400 cu. ft./hr.

The temperature of the gas ranges from 10,000 deg. F. (5,536 deg. C.) and upwards, at which it is partly ionized to create a "plasma" or mixture consisting of free electrons, positive ions and neutral atoms. The gas is confined in a suitably-designed cutting torch, and when the arc is struck, the gas mixture becomes fully oxidized and attains a temperature of 20,000–50,000 deg. F. (*circa* 11,000–25,000 deg. C.). Electrical power used is d.c., rated at about 400 volts in open circuit, 200 under load, with output of 200 k.w.

The tungsten electrode is contained in a ceramic chamber with a double helix gas-channel and a cooling-water inlet. The plasma-forming gas passes through the gas inlets. The electrode is recessed into a nozzle provided with a small orifice, and a spark at high frequency creates an arc between electrode and nozzle, both water-cooled. This pilot arc is then cut off and the cutting done by the external arc. The nozzle of the torch is of copper and held $\frac{1}{4}$–$\frac{5}{8}$ in. from the work-surface.

The process is economical for straight cuts in mild steel in large amounts, up to 2 in. thick, and is believed to be capable of considerable development as an alternative to other machining operations, such as turning, milling and planing.

Plastic Foam Grinding Wheel In **abrasive belt grinding** (q.v.), a contact wheel made of polyurethane, used for the fine polishing of extreme contours. It is exceptionally soft and highly flexible.

Platen The work-table of a planing machine which carries the work to and from the cutting tool. The name is derived from the platen of a printing machine.

Plug Chamfer Taps The type of solid tap having a plug chamfer as shown in Fig 33. This is probably the most popular form, and gives a somewhat higher output per hour than a **taper chamfer tap** (q.v.) It can be admitted into the hole without particular

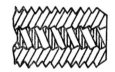

Fig. 33   Plug chamfer tap

difficulty and, assuming enough clearance is available, is particularly serviceable in tapping blind holes. See **Tapping.** The chamfer eases the entry of the tap into the hole. The angle is about 10 deg. or more for tapping stainless steel, and should not be unduly prolonged (3–5 threads for complete holes, less for blind holes).

Plug Damper In boring operations vibration and chatter are sometimes minimized by a plug damper made of a heavy material inserted in the hole of an auxiliary spindle, but being of lesser diameter than the spindle hole by about 0·002–0·003 in., it floats. Should the spindle begin to vibrate, the plug holds fast by inertia, and in consequence the surrounding air is made to flow from side to side, i.e. from the left-hand to the right-hand side of the plug, or *vice versa*. This flux helps to disperse and cut down the vibration (which the plug also withstands) and in consequence the vibrations of the spindle quickly fade out. Heavy damper-plug materials give more rapid damping-down, and for this reason a heavy tungsten alloy is often used and the plug located at the point of maximum potential vibration, if known, usually at the free end of the spindle. Dampers are less effective in proportion as the spindle becomes longer.

Plug Gauges Gauges separate from the tool used in honing, etc.,

PLUNGE CUTTING

and of the same diameter as the required hole. The gauge is presented to the hole at the end opposite to the tool entry at each stroke of the honing machine. Until the hole is correct to size to within 0·002 in. the gauge will not be accepted. As soon as it is, however, the automatic honing cycle is ended by suitable mechanisms. The gauges themselves, sometimes finished to 1–2 micro-in. by ring lapping, check the dimensions of holes or internal threads, and may be plain, screwed, parallel or tapered. If the GO and NOT GO dimensions are embodied in one and the same plug, the gauge is termed "progressive".

The gauges are sometimes in the form of a ring for checking large bores and plug gauges may be reversible so that when one end is worn, the other can be used.

Plunge Cutting A method of grooving parts to close limits of size with an accurately-positioned brazed carbide tool. The cutting-edge of the tool may be honed to reduce formation of craters and abrasion. Heat-resisting steels are sometimes turned by this method.

Plunge Grinding In **electrochemical grinding** (q.v.) the operation in which an abrasive wheel reciprocates in the plane of grinding to ensure uniform wear of the wheel, the work being fed against the wheel. The feed is automatic for large quantities, manual for tools and small quantities. Fewer wheel truings are required, and the finish obtained is superior to that given by other methods. This is the most rapid method of grinding for this particular machining process.

In the cylindrical grinding of splined shafts the same operation is carried out with an angular wheel about 0·021 in. from the pilot-end diameter. Pinions can also be plunge-ground. Some operations in electrochemical discharge grinding are also carried out.

Pneumatic Drum Grinding Wheels In abrasive-belt grinding, contact wheels of inflated rubber used in grinding and polishing where uniform finishes are required and contours have to be ground. The hardness of the wheel is controlled by the air-pressure in the wheel, which can therefore be regulated to conform to exceptional contours.

Point Angles See **Tool Angles.**

Pole Lathe An old-fashioned type of lathe. A foot-treadle was used to bring down the end of an elastic pole located above in a horizontal position. The pole carried a cord which, suspended from it, revolved the work and moved it towards the cutting-tool. By removing the foot from the treadle, the operator revolved the work in the reverse direction. The drawback of this lathe was the loss of time involved in the reverse motion, which both withdrew the tool and stopped the cutting action. Some pole lathes are still found in China and other oriental lands.

Polishing An operation involving the use of abrasives to finish work previously ground. It is subdivisible into roughing, dry-fining and greasing and finishing. The polishing wheel conforms to some extent to surfaces of intricate contour, and is different in density from a grinding wheel. There are many types of wheel for different work, including woven cotton fabric, quilted discs, hide, sheepskin, wool, leather, etc. Some special wheels are the pneumatic drum and plastic foam types (q.v.). There is also a type with flexible x-shaped serrations of rubber used when intricate contours have to be polished. It has a durometer hardness of 30–60, being softer at the working surface than at the hub.

Aluminium oxide is the abrasive most generally used, but jewellers' rouge is also extensively applied to the final stages of fine polishing. Polishing wheel speeds range from 5,000–6,000 surface ft./ min. on normal work. Higher speeds develop excessive frictional heat, which damages the wheel surface. Glued abrasives are not used for mirror-finishes. Wheel pressure is held constant and not speeded up, as this burns and discolours the surface, besides distorting and giving a poor finish to the work. Wheels are ventilated by holes to reduce overheating.

A mandrel and pulley, driven at high speed, can be used for light work. A fine polishing process for finishing metallic surfaces is **buffing** (q.v.), which includes also the elimination of surface flaws from the work, i.e. those impairing the final appearance. The operation provides the mirror-like brilliance and the smoothness of surface required, as well as displaying the inherent tint of the metal. The wheel is different from that used in polishing, being built up of different materials glued or stitched together in sections.

TABLE XII

Grain Sizes for Polishing

Glue per cent (by wt.)	24–36	40–54	60–70	80–90	100–120	150–180	220–240
Grain Size	50	45	40	35	33	30	25
Water per cent (by wt.)	50	55	60	65	67	70	75

The wheel material is of bleached or unbleached linen, flannel or cotton twill. The operation is a rubbing rather than a grinding one. The buffing compound is a mixture of abrasive and grease smeared on to the contact surface of the wheel: chromic oxide + mutton fat is a typical example. Chromium oxide compounds and aluminium oxide compounds are used for steel, aluminium oxide for steel sheets, chromic oxide for stainless and chromium steels. Peroxide of iron compounds known as *crocus* are rarely used today.

Table XII shows the grain-size of a buffing-abrasive when the sections are glued together. The *buffing mop* is a cylinder of cotton material made by forcing discs of the material hydraulically on to a shaft and truing them up on the faces to achieve uniformity. The face is coated with glue and receives two coats, with drying after each coat. The mop is then diamond-dressed, finally coated, dried, trued and put into service. This type of mop is used in buffing the stainless steels.

Polishing is done in band, vertical band, dome and other machines.

Poppet An alternative name for the headstock of a lathe whose rotating mandrel revolves and carries the work between centres.

Positive Rake The rake given to the cutting-edge of a tool used in a lathe, planer or shaper, and such that the cutting-edge is in advance of the cutter-axis and cuts with a shearing action.

Power Hacksaws See **Hacksaws.**

Precision Boring Machines A type of machines used in boring to fine limits (0·00001 in.). In one type the spindle and work are carried by a reciprocating work-table, and in another the work is held in a fixture located on a fixed bridge, the spindle being carried by a reciprocating table. In both types the work may be placed on the spindle and revolved, the tool placed on either work-table or fixed bridge and remaining stationary. These machines are specially devoted to the production of tools.

Precision Threading Machines See **Threading.** Machines with mechanisms for precise control of the lead.

Pre-finished Steel See **Pre-machined Steel.**

Pre-machined Steel Steel of tool type, previously-machined instead of being supplied in black bar form. It enables marking-out operations to be begun at once, reduces machining time considerably, and also reduces labour in the tool-room. Steels often supplied in this way include tool and die steels in lengths of about 4 ft., ground all over to $+0·015-0·025$ in.

Prismatic Tools See **Dovetail Tools.**

Profile Cutters Cutters whose cutting-edges take the form of the cross-section to be given to the work. They differ from form cutters in that they are ground on the narrow surface to the rear of the cutting edge or what is sometimes called the **primary clearance,** and are used in a milling machine, known as a **profiler,** which is highly efficient, of reasonable cost, and extremely accurate. Speed depends on the number of parts to be profiled, being highest for numbers from 500 and upwards or for small quantities from 1–5 only. The working expense is moderate, the form block being inexpensive and the set-up cost also small. (See **Profile Milling Cutters.**)

Profile Grinding In **electrochemical discharge grinding** (q.v.), a means of reproducing a master. There are various techniques:

For single-point tools without flat vertical or horizontal surfaces, a plastic sheet is held in contact with the wheel peak to remove the surplus of electrolyte. In grinding multiple-form profiles without flat vertical or horizontal surfaces, graphite scrapers are used in conjunction with a profile-grinding machine operated by d.c., giving suitable pressures and current densities, without direct contact between wheel and work. For intricate contours, both curved and with flat horizontal or vertical surfaces, compressed air is forced through the central hole of a graphite scraper and regulates the way in which the electrolyte is distributed, especially on those surfaces that are vertical or horizontal. The air-pressure is 40–60 lb./sq. in., and the wheel is separated from the scraper by a gap of 0·002–0·007 in. The scraper exerts on the wheel a force of about 15–30 lb./sq. in. The compressed air is forced through a slot 0·08 in. in width scross the scraper.

In other types of grinding, such as cylindrical grinding, the work is done between centres, using suitable abrasive wheels, but not compressed air. Such parts as round stepped shafts may be ground in this way as long as they are unlikely to be deflected or distorted by the wheel pressure; other similar operations demand that the wheels be of adequate width and output. Case-hardened steel spline shafts have been finished by grinding, while others have been profile-ground, undercut and radiused. Crankpins are also cylindrically profile-ground to finish their diameters and give radius to a shoulder. The wheels used are mostly of aluminium oxide, large wheels being applied (36–42 in. dia.) because they give a greater number of pieces ground with fewer changes of wheel. They are contoured to suit the work. The speed of the wheel is about 590 r.p.m.

Profile Lathes Lathes specially designed and applied to the turning of internal and external forms. They are versatile and suitable for mass production. The pre-established form is reproduced on a revolving component by means of a special attachment either affixed to the machine tool or forming an integral part of it. The cutting-tool used to produce the requisite contour is held in a special appliance and guided along a path according to the contour of the former or copying attachment. The former is normally a slot generated so that the tool travels at 90 deg. to the work-axis owing to the action of a roller fixed to the toolholder. The saddle on which this attachment is carried travels longitudinally, and the cutting tool follows the slot path. See also **Lathes.**

Profile Milling A machine for diesinking, provided with a profiling mechanism and duplicating components with a complicated form when large numbers are required. Jigs are often used in the various milling operations, but the cutter, as in **profiling lathes** (q.v.), is movable and guided in its path by a former having the desired contour, a pin located at a specific distance from the tool enabling it to trace the outline of the master profile. This is often achieved by moving slide and pin laterally as the work-table travels longitudinally. Alternatively, the type of work-table capable of moving in either direction is used.

Sometimes the former has an auxiliary plate to modify the guiding surface according to the class of work, or the profile control may be located straight on to the work.

The machine can be used for duplicating dies and moulds in the type of machine having a vertical spindle for millings, another for tracing and a work-table able to move horizontally in whatever direction is desired. There is also profile milling in which the model is from three to five times as large as the work, the machine having graduations to indicate the control required for the necessary reduction. This is known as profiling by **pantograph** (q.v.), the master being an expanded copy of the die to be produced. Some pantograph millers use a two-dimensional master, others have a tool that travels in three directions. In the two-dimensional pantographs the third dimension involves raising the work-table.

Profile Milling Cutters Milling-cutters whose teeth have been profile-cut or sharpened to give a tooth form capable of reproducing intricate curves blending into angles, narrow projections, narrow recesses, etc., and held to limits of considerable strictness. The cutters are shaped and relieved by form-tools, and are widely used in the automobile and aircraft industries. They are often tipped with tungsten carbide.

Pull Broaches (Fig. 34) Broaches specially used for internal work. See **Broaching.** The broaching tool is fixed to a drawhead, itself connected to a screw turning in a nut. The movement of the nut gives either a forward or a backward motion for the necessary pull

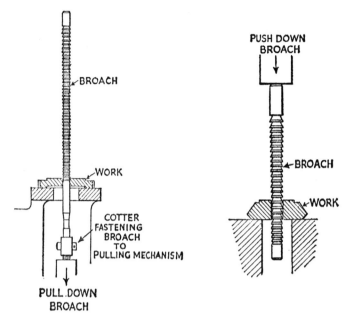

Figs. 34 and 35   Pull broach and push broach

through the work. The pull-length is controlled by adjustable tappets or small pins, which bring the movement to a stop at the right point. Two or more broaches may be pulled by the same machine. In some broaching machines a rack and pinion is used instead of the drawhead, and this changes the rotary motion into a reciprocating motion. A **pull broach** is longer than a push broach.

Pulley Taps In **tapping** (q.v.), that form of taps having identical dimensions with hand-operated taps, their shanks being equal to the principal thread in diameter, but considerably longer than hand-tap shanks. They are used in producing threads in setscrews or oil-cups in the hubs of pulley wheels, and the unusually large diameter of the shanks guides them into the pulley circumference holes, entry into which maintains the taps in alignment.

Push Broaches (Fig. 35) Broaches for internal broaching, but short and sturdy to ensure rigidity and resistance to the severe stresses set up. This shortness and stiffness partly restrict the uses to which they are put. They are usually solid tools with teeth integral to the body or else inserted and renewable, being then locked by suitable mechanism to the broach body. Some broaches of push type

QUADRANT

are pushed through the work in a progressive series of increasing dimensions to ensure accurate finishing of the hole.

The machine used is often a normal vertical press as for punching, or a manually-worked mandrel press for short runs on light work. In the modern shops, however, hydraulic machines are used. In these an electrically-driven pump situated over the press supplies a pressure of 800–1,000 lb./sq. in. to the pistons of a cylinder, which develops a ram or plunger stroke regulated by a valve and lever. The length of the stroke depends on that of the broach, which again is controlled by the number of teeth needed to cut the metal of the work. Hydraulic broaching machines produce holes broached with precision and finishing to fine tolerances.

The machine operates at up to 20–30 ft./min. and the reverse, non-cutting stroke is performed at double that speed or more to economize in time between strokes.

These broaches are mostly used for bringing the holes of bushes and similar parts to precise dimensions. Some, however, are used for burnishing previously drilled, bored or roughly broached holes, or to correct the dimensions and form of hollow bushes or other cylindrical work. They are built up of polished rings of round cross-section which have no cutting edges. Their action compresses the internal surface of the work, flattening out all inequalities and developing a typical and efficient surface-burnishing. The rings are often of solid tungsten carbide, and are used in most instances for non-ferrous alloys.

Yet another application is to blind holes or multiple-station broaching requiring a number of short broaches in place of one long one, to minimize operating time on a specific job.

Q

Quadrant Normally, the segment of a circle bounded by two radii at 90 deg. to each other and by an arc of the circumference; but in machining, the term is commonly applied to the swing (q.v.) plate holding the change-gears in the feed-train of a lathe, or the top portion of a shaper toolbox swinging through an arc of a circle and used in shaping a hollow circular part.

Quick-return Mechanism The means adopted for causing a work-table or other member of a machine tool to run faster on the return than on the cutting stroke. It is mostly embodied in planers, shapers, slotters, etc., and sometimes in lathes with a quick-return slide-rest.

Quill In connection with **lathes,** this refers to a type of auxiliary shaft or spindle employed for holding and rotating work needing exceptional accuracy of hole-position. The spindle rotates on its own rest or bearing-surface, and this rest is situated on the bed of the lathe before the headstock. There are two types of quill, the chuck and the face-plate, but in addition special attachments can be mounted on the spindle end. The quill is not allowed to ride on the internal shaft of the lathe.

R

Rack-feeds A system of fixing power-tools in battery formation for batch production.

Rack Machining That form of machining embodying a rack having from 3–5 straight teeth and reciprocating parallel to the axis of a spur gear or to the helix angle of a helical gear. The cutting action resembles that of a **shaper** (q.v.) and the work is rotated in synchronism with the cutter strokes. At the same time the tool is advanced, and in this manner the correct involute curve of the gear tooth is generated. The technique, less costly than **hobbing** (q.v.), is specially advantageous for machining gears of large diameter or coarse pitch, or these combined, even gears with a $\frac{3}{8}$ in. diametral pitch being machined in this way. The operation may be carried out in many machine tools for gear-cutting. The rack teeth are themselves cut by milling and shaping machines (q.v.), using the conventional types, but there are also special rack-cutting machines for use where required output is large.

Rack Rolling A flat bar provided with teeth of gear form cut straight across is often used for rolling helical splines, the tooth-form being pressed into them by mechanical means. The work is held between two racks or flat dies of the desired tooth-forms and the racks are traversed simultaneously in opposite directions. This traverse causes the work to rotate, and the teeth at each pass bite steadily deeper into the material. This in turn sets up a rearward extrusion of the material to constitute the spline teeth, which may be down to $\frac{16}{32}$ in. diametral pitch, but must not exceed 3 in. The rack presses about 0·003 in. into the material for each tooth, assuming this material to be metallic. The teeth have straight sides and produce conventional involute contours. The depth of their bite depends on the spline-tooth pitch and the hardness of the material.

In rolling helical gear wheels or those of herringbone type the teeth are inclined. Typical applications of rack rolling are to axle shafts, rear axle driving-pinions, transmission shafts, speedometer and controlling worm wheels, steering shafts, drive-line components, torsion bars and starters. It produces splines whose contours are more accurate than those made by alternative types of rolling. Racks up to 48 in. long are used for the operation, employing machines that incorporate a set of racks for rolling and another for synchronizing. The standard rack sizes are 24, 36 and 48 in.

Rack Shaving A method of **shaving** (q.v.) in which the rack is given a reciprocating motion underneath a gear wheel, the feed-in occurring at the completion of each stroke. The largest diameter of gear produced in this way is 6 in.

Rack Tooth Worm Abrasive Wheels A method of generating gears by means of a rack tooth worm whose teeth are **crush trued** (q.v.) into the proper form. The wheel is carried on an arbour and vertically held, or located in a fixture between the machine centres. If helical teeth are to be ground, the machine-slide carrying the work is at an angle identical with that of the gear helix. The work is rotated by an electric motor synchronous with the abrasive wheel motor. As the abrasive wheel traverses axially over the gear wheel face, the teeth are generated.

The abrasive wheel is carefully dressed to give the desired profile, and the synchronization of gear rotation with worm rotation must be strictly controlled. The abrasive wheels used are about 2 in. wide, and are soft to ensure that the temper is not drawn from the gear teeth.

Radial-arm Contour-bandsaws Sawing machines suitable for bandsawing work of considerable mass and weight. (See **Contour Sawing**.) The machine has three principal members, two movable, the other fixed. The moving members, an intermediate radial arm and a cutting yoke, allow the edge of the saw frame to move inside the specified area, the work remaining immobile and supported by a work-table capable of being elevated or depressed.

Radial Chasers Tools used in cutting threads in aluminium or free-cutting brass, etc. They operate at a previously-determined distance from the work-centreline, termed the point height. There are straight and radial chasers, frequently used in self-opening threading dies (q.v.), usually four to a set. The tools are traversed along the thread to rectify the contour of the thread-crests and provide a suitable finish, as well as keeping the thread to the right size. (See **Chasers**.)

Radial Drilling Machine See **Drilling**. A drilling machine in which the drilling head is carried by a radial arm, capable of being vertically adjusted by causing it to pivot around its supporting pillar. Horizontal adjustment is also achieved in the same way. The drilling head progresses along the arm to the requisite drilling position so that the operator does not have to move the part or worktable for each operation. In consequence this machine is particularly suitable for massive and heavy work, in which holes have to be situated at some distance from the edge. The radial arm is readily elevated, lowered and locked in position while this flexibility of movement enables the drill to be quickly placed in the required position without the job having to be re-set.

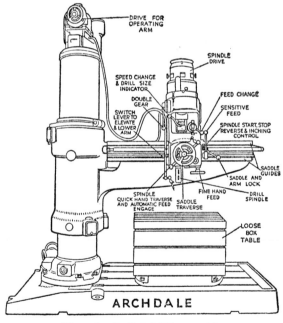

Fig. 36   Radial drilling machine

The drive from motor or belt-cone is transmitted to the gearbox in the drilling head and thence to the spindle, usually through a horizontal shaft running along the arm and splined. The gearbox gives a variety of drill speeds and feeds. The radial arm may be up to 12 ft. or more in length, but is not less than 2 ft. **Numerical control** (q.v.) can be used for automatic drilling with these machine tools as it simplifies the operation, minimizes the possibility of human error, and enables operations to be repeated.

Radial Infeed Thread Rolling Producing threads in a material by causing a revolving cylindrical die or series of dies to travel radially in the direction of the work-centre, the work also rotating. The distinctive feature of this technique is that the least possible axial movement between dies and work takes place during the rolling operation. (See **Thread Rolling**.) The technique enables a thread to be produced close to a shoulder and does not lessen the extent of full thread produced by a special die-face. It is used for thin-walled parts, tubes, stampings and metals so hard that the threads could be formed in no other way. Faulty threads are minimized and threads can be rolled between two cross-sections of more than usual diameter, as in rolling worms on large transmission shafts.

Radial Rake Angles See **Tool Angles.**

Radius Tools Tools designed to produce rounded corners by reproducing their own form in the finished part. They cannot be readily standardized because of their innumerable variations of size and form, but are usually produced from solid bars or by grinding to shape a butt-welded blank of the type shown in Fig. 37.

Radiusing Bore A type of head capable of producing a radius and commonly used to generate a surface described by a conic section, particularly a circle, rotating about a straight line in its own plane. The work is held in a machine tool such as a lathe. The operation is done either manually or with power feed to the head.

Rail Drills Twist drills designed for drilling holes, particularly for the rivets, in railway rails. They are usually of special design, made of high speed steel for the cutting-portion and oil-tempered carbon steel for the shanks, the two portions being butt-welded together; or they may be solid. They have a low helix and a lip-relief angle of 12–14 deg., and drill at about 24–25 surface ft./min., the drill being guided by a bush, and a soluble cutting oil used. These drills are used for ordinary carbon steel rails, but when austenitic manganese steel special points, switches and crossings at those parts of the track subjected to heavy wear are drilled, a specially-designed short and stubby twist drill is used with an extra thick web to withstand

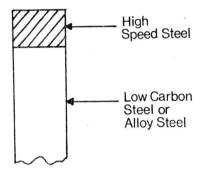

Fig. 37   Butt-welded tool blank

the heavy stresses, and thinned to reduce the bridge or chisel-point to about $\frac{3}{32}$ in. for a 1 in. dia. drill. The rake angle is not appreciably modified nor the chisel point weakened by cutting out excessive backing with the grinding wheel. The drill is of cobalt-tungsten high speed steel, drilling *without a lubricant* at 12–14 ft./min. and with a continuous feed from start to finish of the hole. The drill cone-angle is slightly more acute. The flutes are short in relation to the length and the work is drilled through jigs. As an example of what can be done, a $\frac{5}{8}$ in. drill of this special design drilled 88 holes through $\frac{5}{8}$ in. rolled and quenched austenitic manganese steel (12–14 per cent manganese) before needing to be resharpened, and then only about $\frac{1}{32}$ in. of steel was taken off the point. The holes each took about 2 min. 10 sec. to drill.

These drills are used successfully if the machine is robust and the work firmly held. Both lips cut equally. Lip clearance angle is 7–10 deg. at the periphery and cone-angle 120 deg. Quick regrinding when the drill dulls is essential.

Rake Angles See **Tool Angles.** Angular inclination given to the cutting edges of tools used in machining to obtain the best possible cutting angle.

Raker Set In **contour sawing** (q.v.) a series of three successive teeth set so that one runs somewhat towards the left, the next is central and does not deviate from the straight, the last set to the right, this arrangement being followed through over the entire length of the saw band. The object is to give the saw-back a measure of clearance and allow forms to be cut. It is used primarily for sawing all types of work except thin cross-sections and those in which the cross-section changes abruptly as in tubes, pipes, and constructional steelwork. (See Fig. 50, page 173.)

Ram Stroke See **Shaping**.
Ram Turret-lathe A **lathe** (q.v.) having its turret located on an independent slide or ram and travelling along the axis of the lathe, being carried by a saddle running in the ways and located as near as possible to the work to give the greatest possible support for both ram and turret. Ram and turret move axially on the saddle which, having been set up, is held firmly in position by a suitable mechanism until the work is completed. Its use reduces the effort required by the operator and gives quicker and more convenient handling. On the other hand, this lathe is not well-adapted to make long turning- or boring-cuts, especially with tools held vertically.
Rapid Reverse Planing Machine (See **Planing Machines**.)
Reamers Cutting tools used either to produce an accurate and smooth hole or to enlarge a hole previously formed, as by drilling or coring a casting. Reamers are long, straight or slightly tapered, have cutting flutes running longitudinally, and are either straight or helically cut; the shanks being either straight or Morse-taper. The entire tool is often of solid high speed steel, but in some classes of work it is of adjustable design with inserted high speed steel or tungsten-carbide blades. In many tools of this type the high speed steel cutting portion is butt-welded to an alloy steel shank. The cutting flutes may turn towards either right or left. A right-hand cutting operation requires left-hand helical flutes, but some right-hand cutting taper reamers have right-hand flutes when machining because of their somewhat superior cutting action.

Reamers with teeth on their ends are termed **end reamers** and have right-hand teeth.

Shell reamers are large reamers without shanks, serving the same purposes as solid-shanked reamers, but having a hole, either straight or tapered, running through the centre so that they may be placed

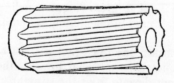

Fig. 38   Shell reamer

on an arbour. The standard hole in these tools has a degree of taper, and their flutes are straight or helical, the most commonly used pattern having helical flutes, taper and bevel lead—this lead, as in nearly all helical reamers, being located at the cutting-end and run-

ROSE-SHELL REAMERS

ning counter to the direction of rotation. It enables the work to be entered more easily. In shell reamers the lead is bevelled.

Rose-shell reamers have straight cutting-edges and the same lead as shell reamers. They enlarge holes from a somewhat smaller

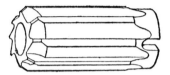

Fig. 39   Rose-shell reamer

diameter and are easy to resharpen, but do not provide so good a finish.

The internal hole to receive the arbour is either parallel or has a $\frac{1}{8}$ in./ft. taper.

Hand reamers have square-ended shanks to receive a suitable driving-spanner and the point has a taper lead, i.e., the cutting edges are slightly tapered to make entry into the hole easier. A **hand bottoming reamer** is used in blind holes, and has square corners but no lead, the end-teeth cutting to a small centre hole. Both these types are hand-operated.

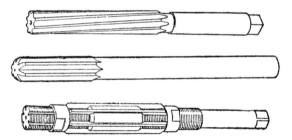

Figs. 40–2   Hand reamer, chucking reamer and adjustable reamer

Chucking reamers give a previously-formed hole accuracy and smooth finish, but also enlarge it to the extent of about 0·005–0·1 in. They have rounded ends, the cutting-flutes being on the circumference of the cylindrical portion.

Adjustable reamers have inserted blades for cutting. In one form the blades are slotted firmly into the body and held in place by

155

REAMING

conical countersunk nuts at each end. In the other the slotting is the same, but the blades are securely clamped by a special device. Other clamping arrangements are possible. The blades in some instances are of high speed steel, in others of mild or alloy steel tipped with tungsten carbide, Stellite or some other hard cutting-alloy, and are about 20 per cent thicker than high speed steel blades. Chatter of the blades must be prevented, so that the clearance angle is usually small.

A **duplex helical reamer** of American design has varying flute-angles to lessen chatter and give a specially smooth finish. There are also various proprietary designs of reamers, fully patented.

The **helical reamer** for conical holes is regarded as better than the straight-fluted reamer, but its design and cutting-flute relief often vary with the maker.

The teeth of reamers are irregularly spaced when chatter has to be prevented. Some makers space the flutes of one half of the tool unevenly, the cutting edges of one half coming diametrically opposite to the other. Others make all the teeth irregular in spacing so that none come opposite to each other. This is claimed to be the better practice as less chatter is set up, but teeth are even in number to improve the cutting action and there is the additional advantage that diameter measurement with a micrometer is facilitated. Helical teeth require more power than straight teeth.

When large holes have to be machined, floating heads are used so that the cutting portion may follow the hole.

Reaming A process whereby a revolving cutting-tool removes only a small thickness of metal to make a round hole more accurate and to give a better finish to the surface. In one sense it is a form of **boring**, and reaming and boring are often used in combination. The operation is mainly confined to holes 0.375–0.250 in. dia. but special reamers will give holes of 0.005 in. dia. The maximum diameter reamed is around two or three inches up to six inches, but for the larger sizes special reamers are needed. The operation is mainly applied to finishing holes in cast iron, steels of Rockwell C_{52} and above, and soft non-ferrous alloys.

Choice of the right reamer for a particular operation is based on the material of the workpiece, its hardness, the diameter of the hole to be reamed, its shape and length, the amount of material to be removed, the method of holding reamer and work, the surface finish and accuracy required, the output, the cost of reaming and upkeep, etc. The types of reamers have been given under **Reamers**.

Reamers are more readily damaged than twist drills, so that the speeds for reaming are usually lower by about one-third than those for drilling the same hole. Speed is governed largely by the material and its hardness, and by the type of material used for the reamer itself. Carbide-tipped reamers are said to cut 3–4 times as fast as high speed steel reamers. Nominal speeds and feeds for cutting carbon and low alloy steels with high speed steel reamers range from 20–65 surface ft./min., and for carbide reamers, 20–260 surface ft./min. For free-machining and low alloy steels the range for high speed steel reamers is 20–80 ft./min., and for carbide reamers from 40–335 ft./min. Speeds and feeds are kept below the point at which chatter begins. Soluble oil is the most commonly used coolant, but for particular materials sulphurized oil, paraffin mixtures, and neutral oil may also be used. Grey iron is best reamed dry, as are some types of brass, but in both instances it is usual to direct a jet of air onto the work, not only to lower the temperature but also to clear chips.

Reaming Machines In most workshops, especially those producing large quantities of identical components, reaming is done in a drilling machine or in a multiple-automatic-turret machine with automatic feed. Manually-held air or electric motor drive is sometimes used for small quantities, or the engine lathe for massive pieces. The form and dimensions of the work are the principal determinants in deciding what machine to use. Drill presses or boring-machines can be used and in some instances even portable drilling machines.

Receiving Gauge A gauge (q.v.), the interior measuring surfaces of which check the uniformity of holes of a specific diameter and form, and also the contour of any mechanically produced material.

Recessing Tools Tools employed for cutting an interior recess or groove in a previously bored hole. The holes are often those produced in castings and forgings, usually at points not readily approached. The tools are mounted in a holder of special design in the hexagon turret or capstan lathe. They are operated by the direct forward motion of the turret, although in some instances a hand lever gives the necessary motion to a toolholder in a slide. These tools are not strong and are therefore made as rigid and large as possible. For aluminium they are given clearance angles of 5–10 deg. and rake angles of 0–5 deg. Larger rake angles cause chatter. **Chip breakers** (q.v.) should be used.

Reference Gauge A reference gauge is made up of a plug gauge with a taper thread and a pair of ring gauges, also taper-threaded. The plug gauge is made to a precise thickness, usually tabulated, and has a gauging notch. One of the two ring gauges has the small end of identical diameter with the small end of the plug gauge, and is flush with the plug gauge at this small end and at the gauging notch, when manually and tightly screwed up. The second ring gauge is also to precise thickness, but is threaded for part of its length. Its diameter at the large end is identical with that of the corresponding edge of the plug gauge. The part of the diameter not threaded is counterbored.

Reinforced Abrasive Grinding Wheels Specially reinforced abrasive grinding wheels of flexible type designed to cut off the unwanted projections such as gates and sprues from castings of complicated form. They are made of laminated sheets of cotton fibre charged with an abrasive, and are applied to heavy grinding, light sanding, tooth sharpening, removal of roughness, and finishing, especially for aluminium castings. See **Grinding**.

Relief Angles Relief angles are given to the cutting-edges of tools by turning, grinding or milling off a portion of the metal behind the tool cutting-edge to give clearance. (See **Tool Angles**.)

Resinoid-bonded Abrasive Grinding Wheels Wheels combining the free-cutting properties of **vitrified wheels** (q.v.) with the safety of the shellac or rubber bond. They are used in swing and floor grinding-machines for removing projections from steel, iron and other castings where rapid metal removal is essential. Lower speeds are used for finishing cams and grinding rolls. The wheels are also used for cutting off metal bars, tubes, valves, etc., and for this work are extremely thin, running at speeds up to 16,000 surface ft./min. in machines of special design. Normally they run at 9,000–9,500 surface ft./min. for other work. Sometimes the resinoid bond is reinforced with fibre.

The wheels are more flexible than other bonded wheels and can therefore be used at relatively high speeds, while they give an excellent finish. They are bonded with a phenolic resin and made in many dimensions. (Phenolic resins constitute the principal group of artificial plastics, and may be moulded, laminated or cast.)

Rest Devices of various kinds for supporting a tool during cutting-operations, as in a lathe, etc. There are also **slide-rests** (q.v.) of numerous types.

Revacycle Cutting A method of gear generation applied to the machining of straight bevel gears up to 10 in. pitch diameter where a high output is required. It is the most rapid method, but is expensive and thus economically justifiable only when mass output of identical gears has to be combined with speedy delivery.

A series of cutters (16, 18, 21 or 25 in number) rotate in a horizontal plane at an even speed. The blades of the cutters radiate from the cutter head and have concave cutting-edges able to generate convex contours on the gear teeth. The work remains stationary, the cutter travelling by actuation of a cam in a straight line across the gear face and parallel to the line of its root. In this way the cutter generates a straight tooth base at the same time as the movement of the cutter and the form of the blades develop the requisite tooth contour. The cutter-blades are progressively of greater length, and this provides the feed. They comprise roughing, part-finishing and finishing blades. Each cutter rotation provides a tooth space, and indexing of the work takes place in the gap between final finishing and initial roughing blades. Consequently independent cutters and settings are needed for each operation.

The cutter has to have a continuous path, and for this reason cutting of gears with front hubs is impracticable. The rate of tooth completion is about 1·8 sec./tooth, and the maximum width of gear face cut is 1·125 in.

Reverse Jaw Chuck See **Chucks.** A type of chuck using dogs (q.v.) capable of being easily reversed, end for end, to enable work to be clamped on the outside or inside diameters by running the jaw right off the screw and turning it round.

Reverse Jaws Chuck jaws for machining external surfaces, being inserted in the part they fasten down.

Revolving Tool Box In certain types of planers used on metals, a box holding those tools which can turn through an angle of 180 deg. to machine equally well in either direction.

Right-angle Boring Head In boring, a type of head frequently applied to reduce chatter and other difficulties experienced with boring-bars of considerable length. It is of particular advantage in machining half-bores. (See **Boring.**)

Right-hand Tools Cutting-tools for use in lathes, etc., which are ground to the appropriate angle on their left-hand side, so as to machine from right to left.

Ring Gauges Precision-ground rings employed in the gauging of finished or partly-finished work to ensure that the dimensions are correct. Typical parts to which they are applied include shafts and external threads. In the United States the standard design is capable of checking measurements from 0·06 to 12·26 in. There are GO and NOT GO gauges, but it is considered better to use a gap gauge (q.v.), for NOT GO work. (See **Gauges.**)

Ring Lapping The least-complicated technique for lapping external surfaces. A cast iron lap in the form of a ring is manually passed and repassed over work held in the chuck of a lathe or head and rotated. The lapping medium is generally a paste used on the surface of the work. The lap is not so long as the work and has adjustable slots if possible.

This operation enables work to be finished to exceptionally fine tolerances, and rectifies lack of roundness where machine-lapping would not suffice. Naturally, being a manual operation, it is costly and slow, so that it is applied only when no other method will produce the desired results. The operation also demands considerable skill. (See **Lapping.**)

Rocker Base That part of a toolpost that fits into the circular seat of the collar, gives support to the tool and enables the tool-point height to be regulated.

Roller Burnishing (See **Burnishing.**) A process whereby better finish and dimensional accuracy are obtained by rolling rather than removing metal, no metal being in fact removed. It enlarges the diameters of holes by 0·0005–0·002 in. with no impairment of the surfaces, but this enlargement is a secondary consideration, the essential needs being precision, fine finish and a surface hardened by work. The method can also be applied to finishing tapered holes, internal and external cylindrical and flat circular surfaces. It is occasionally used in preference to **reaming** (q.v.) and is mainly confined to metals of Rockwell hardness below C40. The work-hardening effect increases surface-hardness to a depth of 0·005–0·030 in.

The machines used are standard types for turning, drilling, boring and chucking, while automatic bar machines can also be used. The process is frequently used after previous boring or reaming operations. The work has to resist the pressure of the rollers, as otherwise distortion of the surface may occur, or the hole may be out of true or of improper configuration. For this reason the walls of the parts to be burnished must be thick enough for the applied pressure, the minimum thickness being about $\frac{1}{16}$ in., but the enlargement for walls as thin as this does not exceed 0·0003 in. on diameter, and thus to achieve the required result the necessary support is given by fixtures.

The operation will not rectify lack of roundness or straightness in the holes. The tools are of varying designs according to the work. Speeds range from 60–150 surface ft./min., feed-range from 0·005–0·200 in./rev. up to 5 in. dia. Copious lubrication is not necessary, and a light spindle oil will suffice, a little being applied to each roller in advance of its working.

Materials that are roller-burnished include aluminium and copper alloys, bronze, magnesium alloys and steels.

Roller Burnishing-peening A method whereby metallic parts are finished on the surface by a combination of **peening** (q.v.) and **roller burnishing**. The peening is executed by hardened rollers revolving about and supported by cams. The rollers are elevated and allowed to fall at a rate of about 200,000 blows/min., to achieve a smoother surface, a rounder or straighter workpiece, and a work-hardening of the surface. The operation is particularly suitable for the internal surfaces of tubes. Various types of tools are used, and speeds of 300–500 surface ft./min. with from 150–250 in./min. feed give good results. Little or no lubricant is required. Copper alloys respond extremely well to this process owing to their lack of hardness and their ductility, which is high, but aluminium alloys, bronze, and magnesium alloys can also be finished.

Roll-forming Gears A method of producing gears at a much faster rate than with ordinary machine cutting. The gears most suited to this method are spur and helical gears. These do not have to be given a further finishing operation after rolling, and the action of the rolls gives the microstructure of the material a finer grain. The tools used comprise a series of rolling dies with the necessary fixtures, the dies forcing the work between them, the teeth being developed by their pressure.

At least 18 teeth are necessary on a spur gear, as any number less than this would not give a good enough result, but helical gears with an adequate helix angle need not have so many. It is usual to grind the gear blanks in advance of rolling to ensure accuracy. The lower the pressure-angle, the greater the power required for pressing and the more sluggish the plastic flow of the metal. A minimum angle is 20 deg. The radius of the fillets rolled is at least 0·01 in. Free-machining or leaded steels are not considered suitable for this work.

Roll Grinder A type of grinding machine of massive construction for particular operations on the surfaces of rolling mill rolls of large diameter, and used to finish them to the dimensions required.

Roll Lapping A method of **lapping** (q.v.) components or surfaces in small numbers, each dealt with individually. The operation is not economical when large quantities are required. Cast iron rolls, one 6 in. and one 3 in. in dia., are coated with an abrasive. Each roll revolves counter to the rotation of the work, the one of greater diameter at a speed of about 180 r.p.m., the other at about 90 r.p.m. Feed is about 2 in./min. A fibre stick holds down the work, being grooved for the purpose, and this stick moves to and fro to reciprocate the piece and regulate its dimensions, but at a slow rate to ensure maximum surface finish and accuracy. Feed can be varied to suit the work.

Not much material is removed, and the lapping is centreless. The precise amount removed depends on the surface finish at the start, and ranges normally from 0·0002–0·0003 in. The roll lapping technique is applied to short runs and has the advantage that the workpiece is rapidly lapped without great expense in setting-up and with the ability to make numerous changes of dimensions. Hard anodized aluminium alloys have been effectively lapped by this method to give a 1 micro-in. finish.

Roll Turners' Tools Tools designed for turning chilled iron or steel rolls, such as heat-treated alloy steel rolls, hard on the surface after continued use over a long period. As these steels and irons generate a great deal of frictional heat at the tool cutting-edges owing to their hardness, the steel has to withstand high temperatures. It must also offer a high degree of resistance to abrasion. The super high speed steels containing tungsten and cobalt in relatively high percentages are mostly used for this work, but are expensive and may not be economical in every instance. They are also less tenacious and ductile than ordinary high speed steels and for this reason must be fully supported by the toolholder to ensure maximum rigidity and freedom from chatter. Some roll-turners' tools of special design have inserted cutting-edges with a body of a lower quality steel.

Root of a screw thread. The lower surface uniting the sides of two neighbouring threads.

Rose Cutter A type of chucking reamer somewhat similar to a **Rose-shell Reamer** (q.v.), but having a semicircular cutting-end provided with flutes, and a shank joined to it by a somewhat narrower neck. It is used for profile-cutting and for making the spherical seats of ball joints, etc. The cutting-end is slightly rounded and the fluted portion has a fancied similarity to the arrangement of petals in a rose blossom.

Rose Reamers See **Reamers**. Reamers that cut on their ends used for enlarging the diameter of holes previously drilled or cored. The reamer body has no flues for cutting, but merely grooves running along the entire length of the body. The grooves clear the chips and enable coolant to be supplied to the cutting-edges. The body-surface is not relieved.

Rose-shell Reamers See **Reamers**.

Rotary Abrader An abrading instrument with longitudinal slots through which abrasive ribbons extend. The inner ends of the ribbons are secured to a rotatable sleeve, which is readily changed and discarded when the ribbons wear out. (Br. Pat. 1,147,915.)

Rotary Double Wheel Grinding Machines Machines to grind work presenting opposed flat surfaces. They embody double wheels, the work being placed between the wheels. The opposing faces are of identical form to ensure that they give each other adequate support and are uniformly finished as the work is introduced between and passes out from the grinding-wheel faces. The machines are for components between $\frac{1}{4}$ and 4 in. in width. They are economical only if the number of parts required compensates for the long time required to set them up and the cost of the work-carrying rotary disc. The machines grind to fine limits of flatness and have considerable grinding efficiency. (See **Grinding**.)

Rotary Filing Filing of metals or other materials by a cylindrical, conical, concave or other form of file. The rotary file is mounted in a drilling machine spindle, but when its location has to be changed by manual control, it may be used in a portable driven machine. These files are often used for finishing punches, dies, models and various other components or tools.

Rotary Gear Shaving A method of **shaving** (q.v.) gears with a revolving cutter whose cutting-edges are a series of serrations on the contour. The gear meshes with the cutter.

Rotary Milling Milling castings or forgings by means of a circular fixture on a vertical milling machine, or on a circular rotating table embodied in a machine of special design. The work has to be easily clamped or released, and the cutting is done continuously with a circular milling-action. (See **Milling**.)

Rotary Planing Sometimes termed **end-milling**, this is the **planing** or **slab milling** (q.v.) of flat surfaces on massive cast or forged parts. The work is done by means of a cutter head, circular in form and of large dimensions, into which are inserted suitable milling-cutters or tools which remove metal progressively as the head rotates.

Rotating Boring Bush A bush revolving in alignment with a boring bar or tool and with a tool-slot in the pilot (q.v.). The helix on the nose of the bar makes contact with the alignment key as it passes into the bar, and causes the slot to revolve in alignment with the tools, so giving support.

Rough-cut The first cut taken by a cutting-tool, to get under the skin of the work. It usually requires more power from the machine than the lighter finishing-cuts, and a slower speed and feed. It is made with a **roughing tool** of the particular form required for the operation and material.

Roughing and Finishing Taps Taps used in combination, the one giving the first cut and bringing the components to rough dimensions, the other producing the desired final thread. The drawback to this method is that as the tools extend beyond the work when the operation finishes they cannot be used for blind holes. See **Tapping**.

Router Bits Specially designed **end mills** (q.v.) with two flutes and a helix angle of 25 deg. to facilitate the clearance of chips when high speed cutting is used. Some routing-cutters, however, have a helix angle of 45 deg. and are used when high accuracy in machining gashes or slots is required. Other cutters, used when sheets in piles have to be routed, have only one flute and embody a pilot (q.v.) to regulate and prevent slight departure from a true path. Their helix angles range from 25–45 deg.

Routing A milling operation in which manual control of the work-feed traces a contour, as in roughing-out the impressions in dies for drop-forging and other purposes. This operation has been largely supplanted by automatic profiling and contour-tracing.

Rubber Bond Grinding Wheels See **Grinding**.

Ruby Lasers In **laser beam machining** (q.v.) a crystalline aluminium oxide or corundum material containing about 1 per cent chromic oxide, non-conductive of electricity and powered by a pulsed light flux, using d.c. and capacitors.

Running Centre Chuck See **Chucks**. A term sometimes used for a chuck that rotates with the lathe, as opposed to the corresponding points in a lathe with dead centres.

S

Saddle In **lathes** (q.v.) a slide or slotted table carried by a bed, cross-rail, arms or alternative surface provided with ways or grooves to provide the path along which it travels. The saddle itself supports other auxiliary slides on which are mounted a **toolpost** (q.v.) holding cutting-tools or a work-table. In **milling machines** (q.v.) the saddle slides on the knee and carries the work table. In **planing machines** (q.v.) and **boring machines** (q.v.) it is usually carried by the cross-rail and itself gives support to the auxiliary slide carrying the toolpost and tools. In **automatic turret-lathes,** the saddle slides axially along the ways of the lathe and is fixed to the horizontally rotating turret, but in ordinary lathes it is a component of a carriage sliding directly on the bed and carrying the cross-slide.

Sapphire Burnishing A burnishing operation for holes up to $\frac{1}{2}$ in. dia. previously produced in parts made by powder metallurgy, it is carried out by sapphire tools mounted in a high-speed lathe or drill press. The component to be burnished is manually forced on to the revolving sapphire tool, a little oil serving as a cutting-fluid.

Sapphire Nozzles In **abrasive jet machining** (q.v.), nozzles of a synthetic sapphire of round form and of dia. ranging from 0·008 to 0·026 in. dia., capable of precision-cutting and having a service-life of about 300 hr. The hardness of sapphire is 1,600–2,200 Knoop.

Sawing The cutting of metal or other materials by hand or machine with a saw in the form of a blade, band or disc with serrations or cutting teeth. The various saws have been grouped into **bandsaws,**

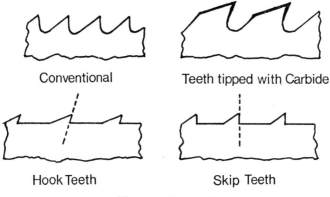

Fig. 43 Saw teeth

circular saws (hot or cold), **hacksaws** and **contour saws**. Hacksaws have straight blades; circular saws are discs with teeth on the circumference; bandsaws have teeth on one edge of a thin ribbon of steel; and contour saws are bandsaws with a much narrower width, usually less than 1 in., and generally $\frac{1}{16}-\frac{3}{8}$ in. wide.

Bandsaws, respectively wide or narrow for either wood or metal, are long strips of suitably hardened, tempered and ground steel formed into a loop by joining the ends. They are placed on the pulleys of bandsawing machines so that they rotate continuously in a single direction only, cutting as they rotate. The work is fed into them. They are hammered to give tension and flatness, the teeth are sharpened and set, and they are used to cut off the heads and runners of castings, ferrous and non-ferrous. They are made in $\frac{1}{2}$–3 in. widths in various gauges and thicknesses, being supplied in long coils or in the required saw lengths.

The teeth are case-hardened (carburized) or chilled for cold steel and cast iron. Those case-carburized have flexible backs. The num-

SAWING

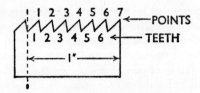

Fig. 44   Bandsaw tooth arrangement

ber of teeth/in. ranges from 6 to 20. Saws with chilled teeth have usually 4 teeth/in. Made in 7 thicknesses from 23 to 20 gauge, they have a thickness governed by the saw width, namely $\frac{1}{4}$–1 in.
Bandsaws with teeth on both edges are more expensive than the standard type.

Bandsaws for sawing wood are of two types—the wide and the narrow or "scroll" (q.v.). The wide range from 3 to 16 in. wide and in gauge from 20 to 12. The **scroll** or narrower type are $\frac{1}{4}$–$2\frac{3}{4}$ in. wide and 22–19 gauge. The highest quality of wood bandsaws are made of a nickel alloy steel for strength and toughness as well as long service-life, but for less exacting work carbon tool steel is used. The wide saws give the best results when run at about 7,000–8,000 ft./min., the scroll saws running best at 4,000–5,000 ft./min. The scroll saws are often made of chromium nickel alloy steel or a good carbon steel, according to the class of work.

Contour Sawing is a variant on bandsawing. The blade, much narrower, is made from an alloy steel, and used in a machine specially designed for the different range of work. The saw is in three standard thicknesses according to the width, namely 0·025, 0·032 and 0·035 in. The back is left soft, but the teeth are hardened to their roots or bases, and cannot be re-set or resharpened once worn down. A wide speed-range is possible, up to 1,500 ft./min. Three contours of teeth are used, raker, wave and straight. (See Fig. 50, page 173.) The number of teeth varies from 6 to 32 per in. Two tempers or hardnesses are obtainable: one for sawing high chromium, high carbon, high alloy and tool steels, stainless steels, low carbon steels, cast iron, hard brass and bronze; the other not so hard, for sawing soft brass, copper, aluminium, thin welded sections, etc.

Hacksaws are used either by hand or in a machine. Hand hacksaws are either full-hardened, have the backs left soft, or are heat-treated to render them flexible like a spring, and have two cutting-edges rather than one. Power-hacksaw blades are either fully-hardened, have spring-tempered backs or are of special design for sawing rails. Most blades are made from tungsten steel and are

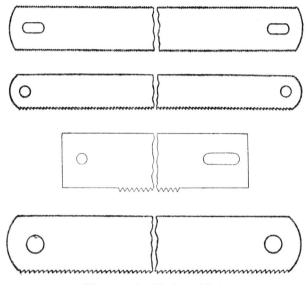

Figs. 45–8   Hacksaw blades

termed high-speed blades. Those of **high speed tool steel** are termed high speed steel blades, and contain about 14 per cent tungsten as distinct from the first type, which have 1·0–1·5 per cent tungsten. They are less efficient than the low tungsten blades for sawing hard tool steel, but properly heat-treated, cut at speeds that would ruin the normal blades in a few seconds. Power-hacksaws should always be lubricated.

Double-edged hacksaw blades are mostly used by plumbers and electrical workers for sawing through soft alloys and cable up to ¾ in. dia. They may also be used on occasion for cutting pipes.

Light power hacksaws have 14–18 teeth/in., the fewer the teeth, the softer the material they will saw. **Rail-cutting blades** have about 8 teeth/in. and a special tooth termed a dog tooth, but they do not cut as fast as other blades because they rapidly wear out. **Heavy power blades** are fully hardened and do not readily fracture.

Sawing Machines Bandsawing machines cut either vertically or horizontally and have a saw band rotating continuously in one direction only. Two large pulleys at top and base in the vertical machine have grooved rims to receive the band, and the pulley revolutions draw the saw teeth successively through the cut. The band passes through a slot in the work-table as it leaves the cut. In modern works the drive is electric, and the work is secured to a movable work-table, operated automatically or manually. The machine base often consists of a trough for coolant in which the bottom pulley revolves. Feed is mechanically varied to suit the work-thickness. The table may be set at an angle to allow bars to be sawn at any desired angle irrespective of their length.

Horizontal bandsawing machines of radial type saw feeding-heads and other undesired projections from large castings. The work-tables may rotate in any desired direction, and in some machines the arm carrying pulleys and blade can be swivelled out of the way to enable an overhead crane to hoist the work into place. Motor-drive is usual, and several speeds are obtained through a multi-speed gearbox.

The **double-cutting bandsawing machine** saws cold iron or steel, such as testpieces and medium-sized castings, and saws two pieces simultaneously with the same band, one cut by the downward-running side, the other by the up-running side. Two self-acting work-tables serve the respective sides, functioning separately, so that parts of varying depth receive the correct feed for each table. The rate of feed is variable but the speed constant. Such a machine has twice the production of the standard, and as all gear wheels for driving are in the upper portion of the machine, the chips do not foul the gears.

Scrapers Thin flat steel blades of varying form used in finishing a machine-surface by hand to give a decorative effect or to modify small defects in flat surfaces that have been planed, bored or milled. They usually have one handle, but two-handled scrapers are also used. The accuracy of the surfaces produced by modern machine tools is such, however, that scrapers are not now often used in production. The tool is presented at almost a horizontal angle so that the cutting-edge removes a thin shaving of the material, the motion being either straight or semi-rotary. Scrapers are of hard steel, and many have been made from old files. They were applied originally to the beds and slides of machine tools, the faces and seatings of engine slide valves, and the curved bearing surfaces of shafts, etc.

Scrapers for Electrochemical Discharge Grinding In profile-grinding by this process, plastic or graphite auxiliary strips, previously formed, against whose periphery the work is passed to spread the electrolyte more evenly over the wheel surface as it enters the gap between control wheel and work. This use of scrapers gives the work an accuracy of ± 0.001 in. in normal production. The scrapers are about ⅜ in. thick and press against the wheel at about 4–7 lb./sq. in. Intricate profiles use a hollow, formed scraper, especially for flat vertical or horizontal surfaces, the pressure used being 15–30 lb./sq. in. and the gap between scraper and wheel about 0·002–0·007 in. Such auxiliaries give a size-variation averaging less than ±0·0005 in. The scrapers are formed on the same profile-grinding machine as the work.

Scratching In grinding-operations the marring of the surfaces produced by undesirable marks, such as narrow grooves, wide irregular scratches, fine helical threads or helices, wavy transverse lines or long, wide indentations. These are caused by a wheel either too hard or too soft, excessive traverse speed, rough wheel-edges, or unfiltered grinding-fluid. (See **Grinding**.)

Screw Machine Drills See **Stub Drills**.

Screw Machines Machines cutting a screw thread on small work, such as screws, pins, etc., made from bar material. The turret-lathe (q.v.) is mostly used. (See **Threading**.) In external screw-cutting the tool form reproduces itself in the work, and is therefore a **form-tool** (q.v.). Thus, the cross-section of the thread-form is identical with that of the tool. A boring tool can be given a vee nose for internal screw-cutting, and will need side clearance to prevent the tool from rubbing on the sides, while the tool cutting-edge will be level with the work-axis. The true setting of the tool depends also on the helix angle of the thread.

The small turret-lathes used have a collet chuck in the spindle and mechanically feed the material through the spindle, but for large quantity production multiple-operation machine tools are employed, usually automatic.

Some milling machines produce screw threads and grinding machines are also used. The various machines form different types of screw, such as those with vee threads, square threads, or special form. In screw-cutting by the lathe a master-screw having the desired thread form and known as the **lead screw** is copied as regards pitch, the profile being given by the cutting tools. Pitch-variation is given by gearing the lead screw to the lathe spindle in the headstock, the headstock spindle providing the drive for the lead screw, to whose end the driven wheel is secured.

Screwing Tackle The generic term for those tools and other objects employed in screw-cutting, such as taps, stocks, dies, etc.

Scribing Marking out the contour of a part to be machined either mechanically or manually to obtain a precise guide in the later machining work. The marking out is done on the rough metal with a **scriber,** a tool made of low carbon steel pointed at one end, with a knife edge at the other. It marks a material to show the location of holes, boundaries, angles, etc.

Scroll Chuck A type of **universal chuck** (q.v.).

Scroll Saws See **Sawing**.

Seaming In **thread-rolling** (q.v.), the flowing of the work-metal up the die flanks more rapidly than at the centre of the thread-contour so that the crest of the thread shows an open seam or fold. This defect is caused by the blank being undersized. The softer and more ductile the work-material, the more pronounced is the seaming. Seams lessen the service life of the threads, particularly when corrosion is encountered.

Second Tap A tapping tool midway between a taper and a plug tap in size.

Sectional Taps Taps having inserted shanks and resembling **shell taps** (q.v.) except that normally shell taps have a longer cutting-section.

Segment Dies In planetary **thread-rolling** (q.v.), dies having an internal radius equal to or a little less than the sum of the rotary die radius and the minor diameter of the threaded portion. The length of the recesses in which the dies are held governs their maximum segment length, and their surfaces are shorter, and on the whole their service-life less, than those of the rotary die. The dies are concave and fixed, being set close to the exterior of the rotary die, which is central and on a fixed axis. The starting end of the segment is adjusted to ensure that the blank is rolled between both dies and traverses the complete arc of the segment die before falling out of the threading area. (See **Thread-rolling**.) The finishing end of the segment gives the necessary thread size, but in some operations this end does little work, as the thread is fully-formed when only halfway through the die. This applies to material rapidly-work-hardening when rolled.

Segmental Grinding Wheel A wheel made up of segments is used in a machine with a horizontal spindle, and is normally large enough to cover the entire work-surface so that a single work-table travel carries out a full cut. The wheels are used for massive pieces when high output and fair accuracy are needed, the work-table travelling at 15–80 ft./min. Such wheels are often used in tool-grinding and are of coarse grit, soft grade and open structure, while they are kept true by dressing-wheels of standard pattern.

Machines fitted with these wheels are slightly more economical in wheel wear than are those using cylinder wheels.

Segmental wheels are also used in vertical-spindle grinding machines. (See **Grinding**.)

SEGMENTAL SAWS

Segmental Saws Circular saws whose teeth are contained in sections by a suitable locking device secured to the saw plate. Each segment or section comprises a suitable number of teeth, and can be readily removed for replacement. Segmental saws are in many instances more convenient and less expensive than saws in which each tooth is individually inserted. (See **Inserted-tooth Saws**.) They are used for cutting metals in the cold state and are of hardened and tempered steel of high tensile strength, the teeth being of high speed steel (q.v.) or tungsten carbide (q.v.). The segments are interchangeable as long as the saw is of corresponding diameter. 3–8 teeth/segment are usual, the number being in accordance with the nature of the work. The carbide-toothed saws are often used to cut non-ferrous alloys at high speed and are brazed to the disc. Fig. 49 shows the tooth forms for roughing and finishing.

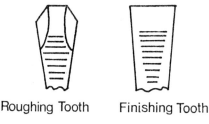

Roughing Tooth   Finishing Tooth

Fig. 49   Segmental saw teeth

Self-aligning Drills Drills so mounted that within certain limits they can align themselves with the hole to be drilled.
Self-centring Chuck See **Universal Chuck**.
Self-opening Threading Dies In threading dies (see **Threading**) a die of either revolving or fixed type for cutting outside threads in cylindrical or tapering surfaces. The revolving dies are used in a drill press, the tool revolving and the work remaining stationary. The die incorporates a yoke which, as soon as it encounters a stop, opens the die by pressure so that the chasing tools withdraw from the work, allowing the die to resume its starting position for the following cycle.

For stationary dies the work revolves, the die itself being fixed, usually in a turret. They open by means of suitable mechanisms at a previously-fixed point, after which the die goes back to its starting position. It is possible to feed in these dies axially as threading proceeds.

Self-opening dies are higher in first cost than the solid type, but the chasers involve less expense in resharpening, so that there is a small overall cost advantage in their use. They are more readily adjusted, and are used whatever the degree of accuracy required within the limits of 1–5. Moreover, it is not necessary to reverse the spindle to withdraw them from the work; they clear their chips more readily than solid dies; and they make it possible to prevent stop lines. Probably the facility with which dies can be removed is the primary advantage in their use, as it not only increases output and saves time, but also minimizes both thread damage from chips and tool wear, as well as prolonging tool life. Smaller thread diameters do, however, render chip-clearance more difficult.

The dies being opened just before the point of full advance, the cutters are withdrawn from the work slowly with a forward rotary motion, and this prevents the surface cut from being marked.

Self-opening dies are also used for threading pipe. They have broadly the same advantages over solid dies as for threading, but in this instance have higher upkeep and first costs and are more likely to be clogged by cutting-fluids that have not been filtered to eliminate swarf.

Serial Taps In **tapping** (q.v.), standard taps whose diameter has been modified so that each following tap proportionately increases the diameter. They are particularly suitable for tapping some of the aged nickel alloys of high strength and hardness.

Serrated Contact Wheels In **abrasive belt grinding** (q.v.), a standard type of rubber wheel having a durometer hardness of 30–50, used in belt grinding or smoothing cut-down projections or superficial blemishes. It produces a rough-to-medium surface and has a long service-life.

Servo-controlled Contour Bandsaw Feed Attachment In hydraulic bandsawing, a control on the feed system that keeps the feed-power unchanged at the previously-chosen value, irrespective of cut-radius variation, work-thickness or hardness, etc. By means of a hand wheel a sprocket is rotated, and its chain rotates the work to follow the contour of the cut. Heavy work is more readily handled, with greater accuracy, while insufficient or excessive feed is prevented, so raising output.

Set of Teeth A term used for the arrangement of the teeth of metal-cutting and other saws of circular, band and hacksaw types. The forms of set include: (i) **straight,** in which one tooth inclines to the right, the next to the left; (ii) **raker,** in which one tooth, called the raker tooth, is straight, the next inclines to the left, and the third to the right: the raker tooth clears the chips and removes swarf from

SHAPING

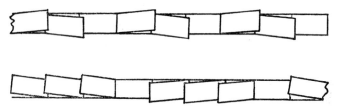

Fig. 50 Raker and G-set contour saw teeth

the cut; (iii) the **G-set,** with one tooth inclined to the right followed by two teeth similarly inclined, then a straight tooth, and then three teeth to the left: this provides the minimum tooth spacing. These forms of set make the cut wider than the saw blade thickness and minimize friction of the blade in the cut. The teeth are re-set after about two or three re-sharpenings. See Fig. 50.

Shaping The machining of metallic surfaces by a machine similar to a planing machine, the difference being that the planer is confined to flat surfaces, and the shaping machine will machine convex or concave forms. It is also easier to operate and has a faster stroke, the essential principle being that in the shaper the tool moves over the work, which, apart from a slight lateral feed, remains stationary, whereas in the planing machine the work moves and the tool is fixed.

The shaper is used for smaller work and quantities than the planer, is more economical and easier to work than the milling machine, costs less in tools and maintenance, and is not so high in first cost.

A typical shaper has three or more speeds in its gearbox, sliding gears operated by a lever working in a gate in the gearbox mounted at the rear of the machine. The maximum stroke is 36 in., and 8–72 strokes/min. can be had. The body contains a ram-driving mechanism, a rocker arm, which can be adjusted as the machine is running, providing a steady thrust over the entire stroke. The tool-head is carried by a graduated swivelling base on the ram, and set to angular variations up to 45 deg. each side of the vertical. Reverse longitudinal feed and push-button control are incorporated.

The ram reciprocates and gives the cutting-motion. The table can be moved vertically and transversely. The toolpost and the arm or ram on which it is mounted give the cut, but for vertical surfaces a vertical feed-motion is available. Inclined surfaces are machined by adjustment of the toolpost. Length of stroke is also adjustable. The ram reverses and runs back quicker on the reverse than when

SHAPING

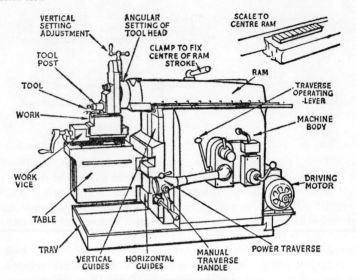

Fig. 51   Shaping machine

cutting, to save time. It also begins and completes the cut slowly, but travels faster in between, so minimizing the stresses when tool and work first meet. Drive may be hydraulic, electric or belt.

The **traversing head shaper** is mainly for heavy forgings and castings, the work being stationary, the ram moving transversely to give the feed. The tool gives both cut and feed motions.

The **universal shaper** has the work-table swivelling on an axis parallel to the ram-movement and with an auxiliary slide for tilting. This can be set at an angle to the line of ram-movement. These machines are used in tool-rooms for machining press tools and dies and work with inclined surfaces needing accuracy in machining.

Shaping is also carried out by *draw-cut shaping machines*, in which the cutting-stroke is towards, not away from, the machine body. The shaping tool is inserted in the toolpost with its cutting-edge towards the body of the machine, and the pulling action makes heavy cuts easier by lessening the chatter of tool or work and the stress on saddle and bearings, while the ram is made more rigid.

Shaping tools have a larger cross-section than ordinary roughing tools used in the lathe, to give them adequate strength. Rake angle and front clearance are smaller than in turning-tools to prevent the digging of the tools into the work or the forcing upwards of the work-table.

The operation is versatile, setting up is quick, and the tools are not expensive, but it is a less efficient machining operation than others. The cutting-stroke is 6–24 in., nothing longer being practicable because accuracy to size falls away as length of cutting stroke increases.

The tools are mostly of high speed steel, carbide tools being for maximum speeds only on work for which they are suitable, and for finishing-cuts when a superior surface finish is required. They are also used in the shaper for machining heat-resisting and other materials presenting difficult problems. The carbide tools must withstand the shock involved and are automatically lifted out of the cut so that they do not come into contact with the work on the return stroke. Single-point cutting tools as used in lathes are not suitable for shaping.

Shaving In multi-operation machining, an operation in which a tool is presented at a tangent to the work and travels either over or under it as circumstances direct. It is usually mounted on the rear cross-slide, the cut being made while the turret tools are also in use and the spindle running forward. The operation finishes work previously rough-machined with a circular form-tool or some other tool cutting on the outside of a component or surface. Tangential shaving is frequently used for finish-turning brass and aluminium, and is sometimes known as *skiving*.

The tools produce an entire form from the solid after the roughing-out, or shave a few thousandths of an inch from the work after a form-tool has been employed. They are used mostly on soft or free-machining materials where a wide, formed surface comprising a range of diameters is to be produced. The tools are usually blocks of steel whose form is generated along their entire length. Travelling over the work, they develop the form from the front cutting-face, the tool-body polishing and burnishing as it moves. Tools of this type are not highly popular, however, as they rub too much and, in regrinding, much steel has to be removed, while the heat of regrinding softens the tool and lessens its cutting efficiency.

Shaving of gears is essentially a finishing technique, used mainly for spur and helical gears, and takes off a thin slice of material from the teeth sides or flanks. It also corrects minor faults in the spacing of the teeth, gives the teeth a superior surface, and may prevent excessive load on their ends, so that they are able to carry a greater load and are more silent when running. Gears having diametral pitches of from 180–2 can be shaved and finished effectively. The gears can be cut in sizes from $\frac{1}{4}$–220 in. pitch diameter.

The shaving tool and gear are at crossed angles. The cutter-teeth are suited to the type of gear being cut. The tool-teeth shave fine cuts from the tooth-contours, being vertically serrated for the purpose. The serration-depth regulates the service-life of the cutter, which is periodically resharpened by re-grinding, a specialized operation, so that in most instances the cutter-regrinding is done by the maker of the tools. The technique adopted is either rack- or rotary-shaving. In rack-shaving the rack moves backwards and forwards below the gear and is fed-in after each stroke. This method is confined to gears of 6 in. dia. and under.

Rotary-shaving uses a cutter much resembling a gear, and here the machine governs the diameter of gear that can be cut. The three types of rotary shaving are (i) the **underpass**, (ii) **modified underpass** and (iii) **traverse**. The modified underpass is the most used owing to its short single cycle, low tool outlay, rapid cutting and good surface-finish. The underpass is used on shouldered or cluster gears and is the fastest technique, but the tool outlay is larger owing to shorter service-life, and the amount of metal taken off is small at each cut, so that the gear must be close to the final dimensions before underpass shaving begins. Gears having a considerably greater width than the cutter are shaved by the transverse techniques which are the slowest, but the tool outlay is not excessive. Speeds range from 275–525 surface ft./min. for carbon and low alloy steels, and 275–650 for free-machining carbon and low alloy steels. **Crown shaving** (q.v.) can be done by any of these three methods. The gear meshes with the shaving cutter.

Carbide tools are much used for shaving gears at high speeds when the conditions are suitable, otherwise a good-quality high speed steel is preferred.

Shear Cutting Gears A technique for forming the teeth of spur, but not helical, gears. The spaces between the teeth are all machined at one time and progressively. The cutting is done at speeds similar to those used in **broaching** (q.v.), and gears up to 20 in. dia. by 6 in. width of face can be cut in this way.

The cutting-head is stationary, the work being forced through it. The shearing tools feed radially into the head by a previously-set distance for each stroke, and then halt. The technique is far from inexpensive, so that it is used only when long runs are required.

Shearing Machines Machines employing shears or shearing-knives for shearing either the straight and free or the irregular edges of sheet metals. They will cope with most materials with high efficiency, speed, accuracy and low cost, as long as shearing is straight. For irregular forms, cut by rotary shears, the minimum radius

SHELL BROACHES

sheared is 1·5 in., there being no maximum radius. In all types of shearing the cost is small. Punching and shearing are often embodied in one and the same machine, the operations being performed one at the top, the other below; or when the machines are of larger size, at opposite ends. The machines may be used for rough-cutting of plate or sheet-edges or for precision-cutting. Some of the machines have quick-return motion (q.v.), but the smaller machines do not incorporate this mechanism. If cuts are not suitable for a punch-press, a saving is often effected by using a **router bit** (q.v.). The machines will cut stainless and other steels, aluminium and its alloys, nickel alloys, magnesium and its alloys, etc., up to a maximum thickness of 0·312 in. for either a single thickness or stacked pieces where the magnesium and aluminium alloys are concerned and cutting is straight, i.e. for free and straight edges.

Rotary machines are best applied to parts numbering not more than 5 at a time. For larger quantities they are less accurate and, for quantities of 500 and upwards, low in speed and economy.

The length and form of the knives depend on the type of work; they resemble scissors in their action.

Shell Broach A **broach** (q.v.) whose body consists of roughing and intermediate sections, and includes an arbour to take a removable shell as well as the shell itself. More than one renewable shell is sometimes used. The tools are for both internal and external work where economy is attained by scrapping fractured or worn sections rather than the entire tool. Unfortunately these broaches do not give the same precision and concentricity as solid broaches, but for some classes of work maintain better alignment because they float.

Shell Drills Drills employed for work similar to that performed by twist drills having three or four flutes. They are inserted in an arbour of a drilling machine and have tapered bores. One arbour will accommodate a wide range of these drills.

Shell End Mills Milling tools having teeth on the circumference and one end, mounted on an arbour. They are **end mills** (q.v.) of somewhat larger dimensions. As contrasted with face-cutters, they have cutting-edges longer than half the cutting-diameter instead of equal to or less than half. They cannot be reversed on their arbours, and therefore cut either left- or right-hand as specified. Cutters bought from stock are usually right-hand.

The tools have no shanks and are therefore more economical than end mills. They give a tolerance on diameter of +0·010 in. and −0·000 in. Most shell end mills have helical teeth. Diameters range from 1 to 6 in.

SHELL REAMERS

Shell Reamers A more economical, larger size of **reamer** (q.v.), but without shanks. Shell reamers fulfil precisely the same function, but have a straight or tapered hole and can be mounted on an arbour. The economy is achieved by the use of less high speed steel or cutting alloy. The standard shell reamer has a tapered hole.

The tools have a tolerance on diameter of 0·0002–0·0007 in., for the smaller sizes, to 0·004–0·0014 for the larger. Either straight or spiral flutes are provided, the most commonly used being the shell reamer with helical flutes and taper and bevel lead for entering the work. A different type is the **rose-shell reamer** (q.v.) which enlarges holes of somewhat smaller diameter, is easy to re-grind, but does not give so good a finish.

Shell reamers are more accurate than **shell end mills** (q.v.) which have no lead and so cannot perform the same operations. Shell reamers, moreover, have cutting-edges longer than their diameter and no teeth on their ends. In use they normally remain stationary, the work rotating as in a turret-lathe, but sometimes both reamers and work rotate. They are more suitable for finishing-work than for taking off large quantities of metal.

Some shell reamers can be expanded, and are made of a fluted body with thin walls and tungsten carbide cutting-edges. The reamer is pressed on to an arbour having a degree of forward taper corresponding to a back taper on the shell, which expands in external diameter as it moves along the arbour. To cause this movement, a sleeve of soft metal is placed over the projecting tip of the arbour and a soft hammer strikes the sleeve until the required expansion is achieved.

Expanding shell reamers are designed primarily to make up for tool wear, and the limits of their expansion are from 0·007–0·020 in. according to their diameter (0·5–2 in.). Expansion is uniform and no grinding work of a corrective kind is necessary. The reamers themselves have a brief cutting-section succeeded by a fluted guiding-section made of hardened steel and leading-in the cutting-section. The larger reamers have flutes running along the entire length, while in addition some have special plugs flush with the end so that blind holes can be reamed.

Shell Taps Shankless high speed steel taps bored along their entire length and mounted on an arbour or other driving device, or grooved to take, say, a shank. They **tap** (q.v.) holes of large diameter and with thread pitch up to 8 to the in. or less when a short tool is needed. In such instances they are more economical than solid taps, but not when their diameters are below 1 in.

Shellac Bond Abrasive Wheels Wheels for grinding machines, bonded with shellac as the matrix, made in moulds, hydraulically pressed, and baked at a low temperature. They do not generate excessive friction and are from $\frac{1}{32}$ in. upwards in thickness. The shellac makes a resilient bond for the abrasive grains, but loses hardness when the temperature rises in operation. A shellac bond is freer-cutting than a rubber wheel and gives heavier cuts without damage from frictional heat. The wheels are particularly useful for cutting off hardened steel, damaged taps, reamers, etc.

Shims Thin plates or wedges of metal used singly or in combination as packing pieces to secure a component to the work-table of a machine tool.

Shoe-type Internal Grinding Machine An arrangement in which the work is rotated by a drive-plate secured to the spindle-nose of an **internal grinding machine** (q.v.), the drive-centre being offset in relation to the work-centre, which seats the work securely on the drive-shoes. The work is axially held against the drive-plate by a pair of clamping-rolls located on arms actuated by pressure. This cuts out chuck-grinding and makes the spindle accuracy less critical. By this method wall thickness is limited in variation to 0·0001 in. and throw-outs are prevented.

Short Flute Helical Point Taps (See **Tapping**.) Taps of solid type modified to thread holes in thin cross-sections such as webs and sheet metal. They have flutes at the helical tip only, and for this reason have a long service-life, but are suitable only for those holes of a depth less than 1 diameter.

Shuttle-type Double Wheel Grinding Machines (See **Grinding**.) A double wheel grinding machine for work having more than about 20 sq. in. of area on each face between the wheels. Such machines will also carry small work, but are then less economical than **rotary carrier machines** (q.v.).

Side Cutting Angle (See **Tool Angles**.)

Side Milling Cutters A type of milling cutter (q.v.) which has teeth on the side, but no helix on the top teeth, and takes light cuts for slots required to have a fine finish of surface.

Side Rake Angle The angle measured at right angles to the cutting tool or the downward slope from the cutting-edge. This is not the true **rake angle** (q.v.). Rake and clearance angles should be the same at all points of the cutting edges. (See **Tool Angles**.) Either top or bottom rake can be given.

Side Relief Angle See Fig. 52 and **Tool Angles**.

Side Tool A cutting tool for finish-machining the internal and external areas of a component, the cutting-edge being so ground that it makes an angle with the tool-axis enabling it to reach the

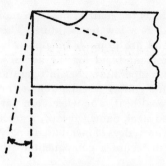

Fig. 52   Side relief angle

sides of the component or those at about 90 deg. angle to the tool-rest.

Silicon Carbide An abrasive made by heating pure glass-sand, finely-ground coke, sawdust and salt in a large electric furnace. At about 2,200 deg. C. (3,990 deg. F.) the silicon of the sand combines with the carbon of the coke to form silicon carbide crystals. Eventually the product is crushed, washed with chemicals, graded and magnetically freed from iron particles. The final product forms the basis of an abrasive grinding-wheel used for: (i) grinding extremely hard tool steels as well as tungsten carbide and other hard materials; (ii) faster cutting in such operations as abrasive jet machining, abrasive belt grinding, honing, lapping, general grinding and ultrasonic machining (q.v.); (iii) work on soft brasses and bronzes, cast and chilled iron, aluminium, copper, marble, granite, leather and non-metallic materials. The substance is also termed **carborundum** and carbon silicide, and is suitably bonded with clay or other materials.

Silicone Dielectric Fluid A fluid passed between electrode and workpiece in **electrical discharge machining** (q.v.). The silicone oils are said to have superior electrical properties rendering machining operations faster and surface-finish finer.

Silver Tungsten Electrodes Sintered and infiltrated silver-tungsten electrodes for machining small slots and holes in **electrical discharge machining** (q.v.), but not for large areas. They are high in cost, and suitable for all materials, while their machinability is fair. Slots in valve parts have been effectively cut using these electrodes.

Sine Bar A tool employed for laying out, setting, testing and other operations of angular work that must be accurate to size. It is made up of a rectangular steel bar, fully hardened, which has its opposite

SINGLE ANGLE MILLING CUTTERS

edges completely parallel and fully square with the neighbouring face. It carries a pair of discs with identical diameter, often referred to as "plugs", and these are mounted somewhat off the longitudinal axis of the bar, their centres being on a line exactly parallel with the edge of the tool. The plugs are inserted in a pair of cylindrical holes whose axes are at right angles to the face of the straight edge. The distance between the centres of the two plugs is usually arranged to be an exact integral number of inches, e.g. 5 or 10, and must be of maximum accuracy.

The method of using the tool is to set it with a single edge at a particular angle to the underside of the angle plate, the angle plate being placed on a surface plate and one of the sine bar plugs reposing on this surface plate. Slip gauges are then inserted under the other pin.

If L is the length between the centres of the plugs and θ the angle of inclination of the straight edge, the desired height of the gauges is $L \sin \theta$, and this can be determined by consultation with sine tables and suitable multiplication.

Single Angle Milling Cutters See **Milling Cutters.**

Single Chaser Threading (See **Threading.**) Cutting threads with a single chasing tool, the work revolving on centres with no die or holder enclosing the work. The operation is also used when multiple-chaser holders cannot be used for lack of tool-space. A standard toolholder is firmly held, the holder opening and closing automatically. As compared to single-point tool threading, it saves time, lengthens tool service-life and, if a flat bevel angle is used, more threads can be involved in the removal of metal.

Single-cutter Trepanning Heads Cutter heads used in **trepanning** (q.v.). They pilot themselves and are carried and directed by wearing-pads mounted at about 90 and 180 deg. to the rear of the cutter.

Single Flute Countersinking Tools Tools for **countersinking.** especially for light work, using a portable apparatus. They cut much smaller holes than tools with multiple flutes, and are specially advantageous when chatter is likely to be experienced. The countersunk hole must be at least 10 per cent of the tool-diameter. These tools are used also when the hole to be countersunk will not accept a multi-flute tool.

Single Plate Lapping Lapping in machines having a single revolving cast iron, or alloy bonded-abrasive plate or lap. The laps are normally of cast iron, but for particular applications may be of copper or non-ferrous ductile alloys; bonded-abrasive laps are also used. The machines function much as a grinding wheel and employ

SINGLE-POINT CUTTING-TOOLS

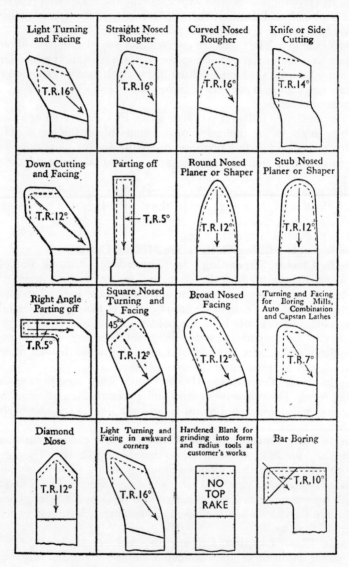

Fig. 53 Typical shapes of butt-welded

SINGLE-POINT CUTTING-TOOLS

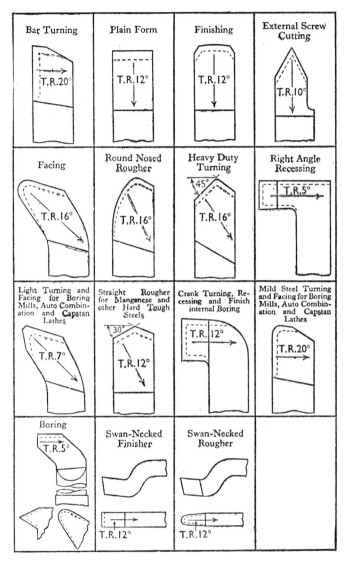

single-point cutting-tools

a loose abrasive such as diamond powder. Except in hand lapping, the work is mounted on carrier-rings which maintain them in position on the lap and rotate as the lap turns, giving the work a cycloidal movement. The pressure-plate and the lap are lapped together to keep the work to fine limits of thickness, flatness or parallelism. Great care is necessary to ensure that the lap remains flat.

Automatic machines of this type feed, eject and gauge the work, while compensating for abrasive wear. They also allow the abrasive to be changed to suit the work. Machines of planetary type have automatic timing-devices for large quantity production. (See **Lapping**.)

Single-point Cutting-tools Also termed **Single-edge Cutting-tools.** Tools with one cutting-edge only located at the end and not always of the same form. Different machining operations require different cutting-edge forms, and even when the operations themselves are similar, special conditions may demand a variation in tool-shape (Fig. 53, pages 182–3)

The tools are divisible into four main types:

Solid tools having both shank and cutting-edge made from one homogeneous material, the cutting-end being first forged to shape, then hardened and finish-ground. They are usually made from high speed, alloy, and less often carbon tool steels.

Butt-welded tools in which a solid end of high speed steel is electrically butt-welded to a shank of low carbon or alloy steel, and afterwards ground to form.

Tipped tools in which a small tip of high speed steel or a cutting-alloy such as tungsten carbide, Stellite or a ceramic material is moulded or cast to form, brazed to a low carbon or alloy steel shank, then finish-ground to the exact size and shape required; examples are: **cermets** (q.v.) and **atomic hydrogen welded high speed cutting tools,** a British patent in which the tips are welded to the shanks by a special process known as Athyweld.

Toolholder bits, called in the United States **tool bits**, small pieces of high speed steel ground to shape by the user and used without shanks in a special toolholder.

The distinctions between these groups of tools are important. The tools are not all straight, but may be bent or cranked to bring the cutting-edge to bear on the specific area of the work to be machined. (See also **Tool Angles** and **Diamond Tools**.)

Single Ram Broaching Machine A machine for **broaching** (q.v.) performing only one operation or giving one stroke of the ram at a time.

Single-ribbed Grinding Wheel Sometimes termed a **single-edge**

grinding wheel, used in grinding threads. The wheel-edge is given the thread form and its contour functions in the same way as a **single-point cutting tool** (q.v.) used in the lathe for screwing threads. The wheel diameter goes up to 20 in., so that at full depth the contactual arc is significant. In consequence the wheel is slanted to the helix angle, especially when that amounts to 4 deg. or above.

This type of wheel will rough 0·02–0·04 in. with no overheating of the work. The feed is 0·0015–0·004 in./pass when great precision of finish is required, but for less exacting operations may be 0·004–0·010. As long as the cut-depth is below 0·04 in., a single cut produces the required finished dimensions, but only at a relatively slow speed of $1\frac{1}{2}$–2 surface ft./min. For deeper cuts than this a speed of from 3–4 ft./min. to 6–8 ft./min. is required.

Single-ribbed wheels are for the centreless grinding of screw threads as well as for cylindrical grinding. See **Grinding.**

Single Thread Milling Cutters In milling threads (see **Threading**) a form of milling-cutter employed when coarse pitches are required or when the threads exceed in length the capacity of a cutter of multiple type.

Skin Milling See **Milling.** Milling involving cuts of considerable depth made with **slab cutters** (q.v.), often employed in making aircraft parts in a **plano-miller** (q.v.) in which complex tracer-control is incorporated. By this process cuts as deep as 2 in. have been made in an aircraft wing skin with a cutter having carbide inserts secured to the body by brazing, at a speed of 10,400 surface ft./min. and a feed of 0·006 in./min. The material was an alloy.

Titanium alloys are also skin-milled at roughing-speeds of 20–125 surface ft./min.; 25–165 for finishing with high speed steel milling cutters; and for carbide cutters at speeds of respectively 60–400 and 80–530 surface ft./min. for roughing and finishing respectively. Higher speeds still are obtainable if the carbide cutters are of the **disposable insert** type, in which the tip of tungsten carbide is indexed to the following cutting-edge without taking the tool out of the machine, the tip being scrapped after all its cutting edges (6 or 8) have been employed. In this instance the speeds may be 65–440 surface ft./min. for roughing and 90–585 for finishing, according to the material being cut.

Skip Rib Thread Grinding In contrast to grinding in the ordinary way with a multi-ribbed wheel, having two or more parallel ribs about the circumference, this operation uses a multi-ribbed wheel in which the spaces between the teeth are such that the wheel grinds only on alternate turns of the thread during the initial rotation of the work, so that the thread is finished only on the second rotation.

About 2½ revolutions are necessary for the grinding of a complete thread. See **Thread Grinding**.

Skip Tooth Saw A saw having greater spacing between the teeth so that chips can be more readily cleared. The gullet is not so deep as that of the normal saw tooth so that the teeth are coarser in pitch on a narrow band for **contour sawing** (q.v.). Such teeth are advantageous in sawing soft metals to a considerable depth. The sawing angles of the teeth are largely identical with those of the standard contour saws.

Skiving See **Shaping**.

Slab Miller See **Plano-miller**.

Slab Mills Milling cutters having teeth on the periphery and, on occasion, some side-cutting teeth, mostly used for peripheral milling in horizontal milling machines, the cutters being mounted on an arbour whose axis lies parallel to the work-surface. Some mills have inserted blades embodying **chipbreakers** (q.v.) to produce satisfactory finish and even-out the cutting load. Cut-depths depend on the availability of power, but should not be less than 0·005 in. for hard and difficult materials, or wherever work-hardening is likely to occur. See **Milling**.

Sleeve A hollow cylindrical piece or **quill** (q.v.), in which a tool is inserted; e.g. a twist drill.

Slenderness Ratio The relation of length to minimum gyrational radius, especially in the phenomenon known as "whip", i.e. the tendency of a long thin tool to show a slight bending movement under sudden stress. The ratio should not exceed 20 in tools.

Slide-lathe A lathe in which a slide-rest (q.v.) is incorporated.

Slide-rest An essential part of a lathe of automatic, screw-cutting or other type. It comprises a saddle (q.v.), secured either to the lathe-bed vee-edges, or between the bearers. In addition two slides travel on veed edges, one sliding, the other surfacing. A circular swinging movement enables angular turning to be achieved. Various means of power or hand movement of the slides and saddle are adopted. In the larger slide-lathes, the rest is one with the lathe, and no cutting can be done without it. In smaller lathes it is an attachment used for automatic turning only and, if hand-turning is required, is removed. Such lathe attachments are, however, rarely seen today.

Sliding Ram In a shaping machine the arm or ram driving the cutting tool forward and bringing it back, usually in a horizontal direction; but vertically-acting rams are also found in shaping machines for cutting vertical surfaces. The work is carried by a flat bed in both horizontal and vertical machines. See **Shaping**.

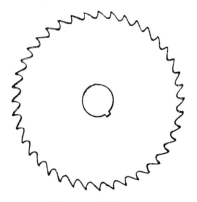

Fig. 54 Slitting saw

Slitting Cutting deep slots in heavy steel sections or cutting off sections of thin metal or other tubes with cutters having teeth on their periphery only, the sides being ground concave to give cutting clearance. They are often termed **Slitting Saws,** some being extremely thin and delicate, so that they have to be handled with great care. The pitch or number of their teeth is governed by the material being slit, a fairly large pitch being used for solid work and a smaller pitch for thin sections or tubing. The saws commonly made measure $2\frac{1}{2}$ in. dia. $\times \frac{1}{32}$ in. thick up to 8 in. dia. $\times \frac{1}{4}$ in. thick. The thicker saws are used for cutting-up tubes or rods into lengths.

These saws or cutters are often used to advantage for sawing copper alloys when a high degree of surface-finish on the cut surfaces is essential.

Steel plate is sometimes slit by means of **bandsawing** (q.v.).

Slotting Cutters Tools for cutting slots or keyways. They have to resist end-pressure, and since their noses are comparatively narrow, they are liable to jump out of the cut and suffer distortion. The clearance is usually given by offsetting the tool, while the top rake angle is made as small as possible to prevent the tool from digging into the work at the start of the cut.

They have cutting teeth on the circumference only, but are much narrower in proportion to their diameter than the standard **plain milling cutters**. The teeth are straight up to $\frac{3}{4}$ in. wide, but helical above this width. The sides of the cutters are ground with a slight concavity for cutting clearance. The tools are used to mill slots accurately and the standard patterns have a tolerance on thickness of 0·001 in. maximum above or below the specified dimensions.

SLOTTING MACHINES

These tools wear first at the tooth corners, and must be resharpened promptly or much grinding will be needed later, while in the worst instances the cutter may prove unserviceable for accurate, well-finished work. They are superior to power-saws for producing slots. Slotting cutters of width above 0·5 in. are usually more economical when inserts made of tungsten carbide are used, but for narrow slots solid high speed steel cutters are better because of the lack of space available and the need for rigidity. The inserts are made by brazing carbide tips to carbon steel, and locking them in place.

Slotting tools are used in the machining of gears and are usually specified by width of point and cut depth. The tool is guided by a template. **Slotting drills** are short drills with straight flutes and no point, used for milling slots or shallow keyways from solid material. They are made of high speed steel, the shanks being either parallel or taper. The flute-length is about $1\frac{1}{2} \times$ diameter. There are two patterns, one having a slot or groove across the end to separate the two cutting lips, the other having a web uniting the lips.

Slotting Machines Machine tools to finish slots in a workpiece when a **planer** or **shaper** cannot be used. An alternative use is the production of flat or curved surfaces more effectively than with a tool having a horizontal motion. The machines resemble shapers (q.v.) but their ram carries the slotting tool vertically at an angle of 90 deg. to the work-table, and is counterbalanced by a massive weight so that less power is required. The work-table, mounted on a saddle, either revolves or travels transversely along it, there being also a motion of the saddle at an angle of 90 deg. to the bed of the slotter. Control is usually automatic today, the tool cutting on the down-stroke towards the table.

Two or more duplicate parts may be slotted together, one being clamped over the next in a stack, the slotting tool then slotting all the parts to the same form.

Magnesium has been slotted with a carbide-tipped circular sawing tool having 72 teeth.

Slow Helix Drills These are alternatively termed **Low Helix Drills** or **Slow Spiral Drills**. They have a relatively thin web and wide flutes to break up the chips and clear them rapidly away. They have straight shanks and can therefore withstand a higher degree of torque, and are particularly suitable for drilling brass and plastics owing to the great rigidity of the drills and the absence of trouble, such as pulling-through when the drill achieves its final penetration or snatching the hole edge and "hanging up". These drills can also be used satisfactorily for producing holes of no great depth in steel, aluminium and magnesium alloys. For brass the point-angle is usually 130 deg., for plastics about 30 deg., with a helix angle of about 15 deg. Plastic drilling is done dry at about 300–400 ft./min. Marble is also drilled with a slow-helix drill having a point-angle of 80 deg.

Sludging Electrolytes In **electro-chemical machining** (q.v.), solutions of sodium chloride and other salts producing an insoluble reaction product or sludge, used as an electrolyte. They are not suitable for machining tungsten carbide or pure titanium by the electro-chemical process.

Slurry, Abrasive In **ultrasonic machining** (q.v.), a slurry made up of water and an abrasive powder circulated by pump and cooled to disperse the heat generated during the process, and to stop the cement holding down the work to glass plates from becoming soft. It also prevents boiling in the gap between work and abrasive and cavitation caused by unsuitable temperatures. The slurry is usually assisted in flowing by holes previously drilled or cored in the work. Holes are occasionally made 6 in. deep with special provision for slurry circulation. With low carbon steel work the slurry is a 30–40 per cent mixture of 180–240 grit boron carbide type and water, cooled to 1·7–4·0 deg. C. (35–40 deg. F.), and is sometimes pumped straight to the centre of the cutting region by way of a small through-hole in the base of the work. Alternatively a 33 per cent boron carbide slurry of 320 grit is used for ceramic materials. The slurry concentration greatly affects the rate at which material is machined off. The higher the slurry concentration up to 40 per cent, the faster the rate of removal. Above this percentage the rate declines sharply, the hardest and swiftest-cutting abrasive being boron carbide, but silicon carbide and aluminium oxide are also used. Boron carbide, however, is longer-lasting though much more expensive. If rust-inhibitors are introduced into the slurry, they must not produce a foam.

Small Hole Drilling or **Microdrilling** Drilling holes of small diameter (0·001–about 0·125 in.). This is a difficult operation demanding considerable care, ever more necessary as drill diameter decreases and hole depth increases. The feed has to be small, but if too fine the drill rubs in the cut and work-hardens the material. Too heavy a feed overloads the drill and causes fracture. The shortest possible drill is used, often best achieved by cutting-down drills of standard length. For holes $\frac{1}{4}$ in. dia. or less, high speed steel **jobbers' twist drills** (q.v.) are mostly used. The hole must be regularly cleared of chips. Holes 0·008–0·010 are often drilled with **spade drills** (q.v.). The drills are embodied in a mandrel with which the drill-point is ground concentric. High speed steels are seldom used for drilling small holes, but some drills for this purpose have a proportion of tungsten in their composition. Holes of 0·35 in. dia. have also been drilled with carbide-tipped drills brazed on to steel taper shanks.

Snagging A **grinding** (q.v.) operation in which the sprues, fins, etc., of a steel casting are removed on a machine for the purpose. A reinforced flexible abrasive wheel is used.

Snap Gauge See **Gauges**. A type of stationary gauge with surfaces for internal measurement, for calipering of diameters, lengths and thicknesses.

Sodium Bicarbonate Abrasive In **abrasive jet machining** (q.v.) an abrasive material employed for specially fine cleaning, such as for potentiometer components. The powder is carefully prevented from exposure to moisture, as it otherwise decomposes.

Sodium Chlorate, Chloride, Hydroxide, and Nitrate Electrolytes used in **electrochemical machining**. See **Sludging Electrolytes**.

Solid Laps Round bars precision-finished to dimensions and used in **lapping** (q.v.), mostly made of cast iron for large work, but of copper or non-ferrous metal for smaller work. They are used by hand for lapping a few irregular sizes of work, and are essential when internal diameters below $\frac{1}{16}$ in. have to be finished, as no other type of lap can achieve this result. Another application is the remedying of irregularities in the bore. Their disadvantage is that they have to be discarded after only a small amount of wear, and the only way in which this loss can be avoided is by regrinding them to smaller dimensions so that they may be used to finish smaller work.

A diamond abrasive is often used, with olive oil as a vehicle.

Solid Reamers See **Reamers**.

Solid Taps Taps all in one piece, usually made of carbon steel, high speed steel or tungsten carbide, for making threads in pipe. The threads progressively and evenly decrease in pitch diameter. The flutes, either straight or helical, form the cutting edges, give adequate clearance of chips, and distribute cutting fluid to the working-area of the tool. They have a degree of chamfer, either taper, plug or bottoming. Straight flutes are more often used than helical because they are simpler to produce and more readily reground, and give a satisfactory result in most operations. Some solid taps are ground undersize to compensate for thickness of deposit in plated work. (See **Tapping**.)

Solid Threading Dies Dies in one piece having cutting edges integral with the tools, but sometimes having chasers that can be removed and replaced. Some dies of this type can be adjusted. The dies are used for cutting outside threads in cylindrical or tapering surfaces. (See **Threading Dies**.)

Solid Tools Lathe tools having no brazed tip or welded cutting-end, as there is with butt-welded or tipped tools (q.v.), and not needing to be held in a toolholder (q.v.), the entire tool from cutting-edge to shank-end being of solid material, whether high speed or other steel.

Soluble Oils See **Cutting Fluids**.

Spade Drills Twist drills (q.v.) of heavier cross-section than standard, mostly applied to drilling large diameter holes from 1–5 in. They are not usually solid drills, a detachable cutting-portion being inserted in a holder, but some, when drilling smaller diameters, are solid and of cemented carbide. They may have either straight or taper shanks, and are of use in drilling deep holes, a central hole being bored through the shank so that a cutting fluid can be forced to the drill point. (See **Oil Tube Twist Drills**.)

The drills withstand greater torque and end-pressure and are more robust, so that chatter is minimized, and there is less likelihood of fracture or **spalled** (q.v.) edges while, as the cutting lips are shorter and have a more pronounced back taper, the drill does not bind in the hole. Moreover, economy is achieved because the detachable bits can be replaced and re-ground, and are suitable for all drilling machines having regulated speed and feed.

Spade drills are not so economical as ordinary twist drills for drilling large numbers of holes, say above 200. Both types of drill cut at about the same rate; however, the spade drills will accept a higher speed, but a smaller feed. They will drill holes of diameter 0·008–0·01 in. and smaller, such as the orifice holes in fuel injection nozzles, for which purpose they are made of a carbon steel containing tungsten. The drills are also known as **pivot drills**.

Spalling The splintering, cracking or flaking off of the cutting edge of a tool in machining. It usually indicates some fault in either the choice of tool material, the conditions of use, or the age and character of the machine tool. Faulty heat-treatment is another potential cause.

Spark Machining A method of intricate machining in a single pass, so that roughing and finishing sections are combined in the same electrode. The electrodes are round, square or rectangular, last a considerable time, and are said to be both accurate and economical. There are also tubes for machining small round holes 0·036–0·189 in. dia. (See **Electrical Discharge Machining**.)

Speeds and Feeds Terms indicating (a) the speed of the cutting

SPEEDS AND FEEDS

tool in relation to the work, and (b) the motion of the tool for each rotation of the work. The speeds are widely different for each type of tool, class of work and material being cut, so that too fast a speed on some materials will blunt or chip the tool, while too slow a speed may greatly increase machining costs. Speeds must, therefore, be regulated to what is economical for a specific operation. No hard and fast rules of general application can be given, so much depending on local conditions, and efficient **cutting fluids** (q.v.) increase speeds. **Free-machining steels** (q.v.) are machinable at higher speeds than normal steels of equivalent type. Higher speeds are possible on annealed than on spheroidized steels, i.e. those whose iron carbide is in small spheroids, as when cold-drawn steels are annealed.

The factors affecting speeds and feeds are: (i) the type and condition of the material machined, which may be tough, full of hard spots, soft or readily torn, easily ignited if too much heat is generated, as with magnesium alloy castings, exceptionally hard, susceptible to heavy work-hardening as with austenitic manganese steel, or brittle; (ii) the operator's skill and experience—the greater these are, the higher the speeds and feeds he will obtain from his machines and the fewer the errors that will reduce these; (iii) the production desired since, if this is high, tools will be run at maximum speeds and feeds even at the cost of other requirements; (iv) the age, type, power and condition of the machine tool, old and worn machines not giving adequate speeds and feeds, or chattering so much that high cutting speeds cannot be attempted, while design may be unsuitable for the operation or power of drive adequate to give good feeds and cut-depth; (v) heat-treatment, which if faulty will render the tools soft or brittle; (vi) cutting-fluids, which may not be of high enough or correct quality for the work, in which case the best speeds and feeds will not be obtained, while some work is best cut without a fluid; (vii) the design and size of the cutting-tools, i.e., when bad design means slower speeds or fracture of the tool—tools too small fracture at the speeds ideal for a properly-designed tool; tools with excessive overhang are slower in cutting because of the risk of fracture through lack of support; and an excessively large tool encourages cuts that set up chatter and necessitate speed reduction; (viii) machining operation, amount of material to be machined off, and cut area, as when the operation is too intricate for high speeds and feeds or is suitable only for high speeds (for example, different feeds and speeds are necessary for milling plane surfaces, gear teeth or screw threads, even in the same material); (ix) effective resharpening of the cutting edges, which do not cut at the best speeds if

SPHERICAL BORING MACHINE

blunt or ground to incorrect angles; (x) surface finish required—if this is unimportant, tools can be run at higher speeds to give a large production but, if important, light cuts at high speed are required, and may necessitate the use of sintered carbide tools, while roughing may need to be done with high speed steel tools and slower speeds; (xi) the economics of the operation, which may decide whether maximum speed with carbide tools is more desirable than a lower speed with high speed or carbon steels, according to the machining cost per piece; (xii) form and design of the work, which may necessitate intermittent cutting at slightly lower speeds or complicated form-tools whose speed must be worked out by experiment; (xiii) ratio of cut-depth to feed, which decides whether heavy cuts with light feeds are to be taken at higher speeds than light cuts with heavier feeds, even though the cut area is identical. Slower speeds are required for light, thin, or springy work. Table VI (page 83) gives a rough guide to feeds and speeds for lathes, planers and shapers, but must be used as a basis for experiment, not as a substitute for it.

Spherical Boring Machine A precision machine tool for close-tolerance contour boring. See **Boring**. The work is secured by clamps to a fixture with an air-operated yoke clamping base attached to a revolving air-chuck. The tolerance is $+0.0000, -0.008$ in.

Spherical Lapping A method of lapping spherical surfaces (see **Lapping**). The laps may be formed to suit the work and used individually in a multiple-spindle machine tool, milling machine or drill-press. The lap is of cast iron and has a taper or straight shank held in a plain or magnetic chuck, the lap being rotated while the work remains stationary. If a dual-spindle machine is used, the work is held and rotated on one spindle while the lap is held by the other and floats, being oscillated to lap the work surface. The laps are concave or convex. Planetary lapping machines are also used, being converted for the purpose. In multiple-spindle machines, which lap continuously, the lap is secured to the counterbore of the already-installed flat lap by brackets, or conditioning rings may be incorporated.

Spindle A long, slender metal bar or rod of small diameter to which cutting tools or drills are attached by way of a suitable arbour or locking mechanism.

Spiral Flute Countersinks See **Countersinking Tools**.
Spiral Flute Reamers See **Reamers**.
Spiral Flute Taps See **Taps**.
Spiral-grooved Abrasive Belt Contact-wheels In **abrasive**

belt grinding, a type of contact wheel whose face is knurled or helically grooved and made of steel having a hardness of Rockwell C52–55, being used, owing to its vigorous abrasive action, for the heavier grinding operations.

Spiral Point Drills See **Drills**.

Spiral Point Taps See **Helical Point Taps**.

Spiral Tooth Bandsaws See **Bandsaws**.

Spline Rolling Forming fine pitch splines on spur and helical gears or on other types of work by rolling the workpieces on cylindrical thread-rolling or rack-rolling machines. Serrations can also be rolled in the same way. (See **Thread Rolling**.) The work is held between centres and the splines may be straight or helical. Two flat racks having the desired form of tooth or serration travel over the work at the same time, but in opposing directions. The rack rotates the workpiece, and under the applied pressure teeth or splines are quickly formed. Grinding may be necessary afterwards to correct any error in dimensions or form. The operation is applied to axle-shafts of automobiles, rear-axle drive-pinions, transmission-shafts, worm-wheels, torsion-bars, etc.

Spoon Bit A boring tool having somewhat the same form as a shell cutter, for cutting a central core of thin wood, paper or cardboard. Its cross-section is crescent in form, and the cutting-edges are ground to a considerable keenness.

Spot-facing A machining operation in which a relatively small area or "spot" is made true by a suitable cutting-tool, usually a **counterbore** (q.v.) or special **spot-facing tool**. When the counterbore is used, the tool takes off sufficient metal to produce a true seat for some component such as a screw-head, or enlarges a hole to receive a cheese-head screw. The difference between counterboring and spot-facing is that the surface to be faced is invariably at an angle of 90 deg. to the hole axis, while the cut depth is not so great. The special spot-facing tools include **back** and **double end** types, the back tools being applied when the down-stroke of the machine-spindle cannot be used because of the contour of the work. In this instance the machine-spindle grips the tool-driver and introduces it into the previously-drilled hole, after which the spot-facing tool is locked on to the driver-shaft, the spindle being revolved and pushed up into the work.

The **double end spot-facing tool** allows both spot-facing and back spot-facing to be carried out on two internal surfaces without taking out the cutter. Cutting fluids are optional according to the work, and dry cutting is often used. Speeds and feeds are not so important as in counterboring, the cuts being shallower. Drilling machines are used for the work, those of the turret type being particularly favoured. Small diameters may be spot-faced on a vertical hand-feed drill-press, and back spot-facing is sometimes done on a horizontal boring mill with a fixed floor-plate.

Spotting Tool A tool designed to produce a **centre-pop** or **starting hole** in the end of a long bar or other piece so that it may be correctly drilled. Usually the cutting point has an included angle of diameter less than that of the twist drill to be used in the eventual operation, so that the drill may align itself truly by first removing material with the corners.

Spread-blade Face-mill Cutting Machining gears with a spread-blade **face-milling cutter** to generate the desired form by means of alternate internal and external blades of a circular cutter, which machine the surfaces of the teeth simultaneously on both sides of the tooth-space. See **Milling Cutters.**

Spring Collet A **collet** (q.v.) of hard steel, open until forced into the cap for which it is designed, springing open later to enable the work to be fed through it. It is used for round and other shapes in a screw-cutting machine.

Spring Temper Bandsaws Bandsaws (q.v.) commonly used for sawing thin sheet metal at speeds of up to 7,500 surface ft./min.

Spring Tool A type of lathe or other cutting-machine tool used in the past and to some extent in the present for finishing a part previously-machined almost to its final dimensions. The tool is curved upwards or downwards behind the cutting edge so that it possesses some elasticity. The tool is not of high accuracy, but sometimes gives a superior surface.

Starting Hole See **Spotting Tool.**

Stay Tap A type of tap (q.v.) of special length (15–36 in.) once much used for screwing the holes for the stay rods in boilers for steam engines. The screwed portion has 12 threads/in., but the lower end is smooth to keep the tap fully concentric with the holes in both internal and external shells. The tap has a reaming section, and is usually driven by an air drill. Some of these taps have a vee thread.

Steady A mechanism to stop long cylindrical work machined in a lathe from wobbling, and either secured to the lathe bed by suitable means or moving with the slide-rest, in which case it is sometimes termed a **following rest** or **back steady rest.**

Steel Ball Burnishing A tool of steel suitable for bearings, pressed through a bore to give it a better surface finish. It is often used for **burnishing** (q.v.) holes in aluminium and its alloys.

Steel Electrodes Electrodes used in **electrical discharge machining** (q.v.) for reverse polarity. They are not tipped, but held in a device, being particularly suitable for short-run dies. They can be easily machined and applied to the mating of punch and holder. As against this, the wear ratio is adequate only for particular combinations of steels, machining requires much more time than for graphite electrodes, and the work-metal is affected to a greater depth than by graphite electrodes.

Stellite A proprietary alloy of cobalt, chromium and tungsten in various combinations according to the purpose. It is non-ferrous, and has great hardness and resistance to wear, being mostly used for cutting at high speeds (300–400 ft./min.) and light feeds. It is more brittle than tool steel, and is therefore not so suitable for operations involving shock or intermittent cutting. Though much used for light cuts at high speeds on steel, it is, on the whole, more suitable for machining cast iron and non-ferrous metals, when it is in the form of tips ground to final shape on special abrasive wheels after they have been brazed into position and lapped or honed to give the keen cutting edge required. Its hardness is not greatly affected by temperatures up to 820 deg. C. (1,500 deg. F.) and it is tougher at red-heat than when cold. The alloy can be electrochemically ground at a current density ranging from 300 to 1,000 amp.sq. in.

A typical composition is 35–80 cobalt, 10–40 chromium, 0–25 tungsten, 0–10 molybdenum, per cent.

Step Drills Drills having more than one diameter, developed by grinding a series of steps on the drill diameter to give a square or angular cutting-lip. They are specially advantageous in those operations needing holes of multiple diameters, and in forming these standard drills can have steps ground in them or their webs thinned. A step drill may also take the form of a combined drill and countersink with a straight shank, short in length. See **Drilling.**

Step Tapering See **Taper Machining.**

Step Taps Tools used in **tapping** (q.v.) threads having identical pitch but of two different diameters, the threads being tapped at the same time. The taps are solid.

Stepwise Displacement The method by which workpieces or tools are displaced stepwise with respect to each other. It employs a mechanical device controlled by a perforated tape. (Br. Pat. 1,147,398.)

Stocking Cutters Cutters mostly used for taking off material by a roughing-cut of some depth, leaving on a sufficient quantity for finish-machining.

Stop Some means of preventing a machine tool or other mechanism from striking either a delicate part of the work or the end of the machine-way, or from causing the motion of the machine to reverse at a fixed point. It is usually a rectangular piece of metal.

Stop Drill A form of drill (q.v.) prevented from penetrating the work beyond a specific point by the provision of a collar suitably placed behind the cutting point.

Stop Lines A trouble experienced in threading work with solid dies. The die travels forward, the cutting portion taking off metal, but as soon as the cycle comes to an end, the cut is sharply arrested. This causes a discontinuity indicated by a line or "step" at the point where the work ceased, and this line cannot be prevented except by discarding solid dies for those of self-opening type (q.v.). The chasers are cutting at the actual moment of arrest, and when reversed over the stop line, the friction resulting from the step abrades the cutting edges and wears them abnormally.

The step created by the cessation of cutting is proportionate in extent to the type of material being threaded, and there is also the possibility with some work that chips may be caught between chaser teeth and threads, with consequent injury to both. The degree of step may be minimized by putting more chasing tools in the die, but this will not entirely remove the stop lines. The problem is particularly difficult when pipe is being threaded. See **Threading**.

Straddle Milling Milling the opposite sides of duplicate parts to ensure parallelism of the surface, two cutters operating simultaneously. The cutters are set on an arbour and separated to the required amount by collars and washers. For this type of **milling** (q.v.) **side mills** having teeth on both circumference and sides are used, often of high speed steel, but for higher production carbide-tipped cutters, with fixtures specially designed to provide greater tool rigidity, enable higher speeds and feeds to be used. Both side-faces of the work can be milled in a single operation.

The cutters have spiral-cut teeth if above $\frac{3}{4}$ in. wide and straight teeth below this width. For heavy work coarse teeth are used. They are made with a tolerance on width of $+0.005$ or -0.00 in., but for some work the width must not be out by as little as 0.001 in. A common name for them is **side and face mills** (q.v.). The teeth may be right- or left-hand. For extreme accuracy, the cutters may be mounted at a distance controlled by slip gauges. See also **Gang Mills**. The operation may machine the top face of the work as well as the opposite edges and the outer edges, and is applied to cutting cast iron, copper alloys, hard and stainless steels and tool steels.

Straight-flute Drills Drills with flutes that are straight and not twisted into a spiral or helix. Often there are only two longitudinal flutes, parallel with the shank axis, and the drills are then specially suitable for drilling brass and non-ferrous alloys of high ductility, but of recent years **slow spiral drills** with wide flutes have been found superior. A taper shank is usually provided and the drills conform in overall size to taper-shank drills of equivalent diameter; but parallel- or straight-shanks can be obtained which also conform in dimensions to those of standard parallel-shank twist drills. The drills of this straight-flute type do not snatch the work nor do they run ahead. In machining aluminium and its alloys, brasses, copper alloys, etc., they sometimes drill holes deeper than six diameters, the machine tool being an automatic bar or chucking machine. (See **Drilling.**)

Straight-flute Reamers Reamers with straight cutting-flutes, suitable for use in drill-presses, turret-lathes and automatic bar or chucking machine tools, and having either straight or taper shanks held in split bushes and collets or by set screws. They usually have a 45 deg. bevel angle and can be applied to most metals. Chucking and rose-shell reamers are typical examples. (See **Reamers.**) Special reamers are also obtainable for keyways, blind holes, etc. A reamer having much longer flutes than the ordinary chucking reamer is sometimes known as a **jobbers' reamer**, and these have flutes double the overall length. They are not so good as spiral reamers for cutting aluminium and its alloys, and if used have margins as narrow as possible to minimize friction, and an even number of blades in pairs opposite to each other, the flute-spacing being variable to reduce chatter and marks on the work.

Straight Grinding-wheels Abrasive grinding-wheels used in generating the teeth of gears, bevelled on each side and travelling backwards and forwards across the circumference of the gear while this is rolling beneath it perpendicularly to the back and forth motion. The wheel reciprocates somewhat like a reciprocating gear cutter, and forms the teeth of spur gears. (See **Grinding.**) Stainless steel is sometimes ground with these wheels at a speed of 5,500–6,500 surface ft./min., using a water-base soluble emulsion, a synthetic solution or a sulphurized oil as the grinding fluid.

Straight Set See **Set of Teeth.**

Straight Shank Drill A drill whose shank does not taper, as do those of the majority of drills, but is entirely cylindrical.

Straight Shank Oil Hole Drills See **Drills.**

Stub Arbour An **arbour** (q.v.) mounted on a vertical milling machine and holding a slotting cutter to which it gives great rigidity when milling slots. A soluble oil cutting fluid is used.

Stub Boring Bar See **Boring Tools.**

Stub Drills Drills having short flutes and short overall length designed to provide the highest possible rigidity without sacrifice of cutting efficiency. They are more usually known as **Screw Machine Drills.** See **Drills.** Mostly they are used in multiple drilling machines or portable drills. The flutes are about half as long as those of the conventional twist drill and can thus drill without the aid of bushes and do not overhang so much. They are of great use when drilling hard or tough materials, such as nickel alloys and stainless steel. High speed steel is usually employed for these tools.

Stub Reamers Tools for reaming (q.v.) in automatic screw machines and floating holders, or wherever a tool of less than normal length is required for reaming. They may be either right- or left-hand for helical cutting.

Stud Setting The forming of holes with a screw thread by tapping components is often followed by the insertion of a stud or studs, screwed into position either manually or mechanically. Special chucks for stud setting are frequently used, and these sometimes accept both studs and nuts, released automatically by opening, without its being necessary to arrest or reverse the machine spindle.

Sub-land Drills Drills used in combination with other tools, such as **reamers** (q.v.) to carry out two or more operations with only a single pass of the tool. The **lands** (q.v.) extend the entire flute length for all diameters. The different diameters up to as many as four are ground if desired, so that drilling-countersinking of flat-topped screws or drilling-counterboring of socket-head screws can be carried out. The drills are of great value and popularity for high output. In general the maximum drill land or diameter should not exceed double the least diameter owing to the variable cutting speeds and feeds used. As they are ground to the proper size along the entire flute length, the cutting lips can be individually ground many times with no harmful effect on the others, and if sharpening is properly and carefully done, the drills give holes of great precision. High speed steel drills are mostly used.

An operation that can also be carried out is the combined drilling-countersinking and spot-facing of bicycle-pedal cranks, using a special drilling and tapping machine with an indexing-table, the drill having two standard flutes and a point.

Sub-surface Milling (See **Milling.**) Milling with an inclined rotary miller employing an auxiliary reservoir in which both tools and work are immersed, the reservoir containing a cutting fluid.

Subzero Profile Milling (See **Milling.**) Producing a profile in a titanium alloy by means of a vertical miller having a hydraulic tracer and carrying out the work at temperatures below 0 deg. C.

SUBZERO QUENCHING

This is claimed to give a much longer service life of the tools and a smoother surface finish on the work. The low temperature is achieved by flooding the work with a solvent+trichlorethylene at −50 deg. C. (−60 deg. F.) or below. The tool is also flooded. The fluid lowers the temperature of the work to between −40 and −57 deg. C. (−40 to −70 deg. F.), and is cooled with dry ice contained in sufficient quantity in a 50 gall. tank or drum kept at −50 deg. C. (−60 deg. F.) and not allowed to exceed this temperature, as the dimensions of the work must still be within the specified limits even after it has returned to room temperature. Use of subzero temperature facilitates milling, gives uniformly curled chips, and reduces tool-wear.

Subzero Quenching A means of preventing a workpiece from warping or distorting when being milled and afterwards heat-treated. The work is quenched after heat-treatment in liquid nitrogen at −195 deg. C. (−320 deg. F.) as a preliminary to machining. Better alignment of holes and economy in time are said to be achieved.

Sulpho-chlorinated Oils See **Cutting Fluids.**

Sulphurized Oil Cutting Fluids See **Cutting Fluids.**

Superfinishing A process of finishing-honing components to produce on them the finest possible surface finish. It is not used to eliminate surface irregularities of the rougher kind, nor to bring work to its precise dimensions, but is suitable for flat, round, concave, convex or other surfaces. Lubricated abrasive stones are used at light speeds and with light pressures on the work surface. The stones are given from three to ten motions, and operate somewhat like a scrubbing brush with cutting edges. The swarf is swilled away by the lubricant, and the stone makes a purely surface contact, minimizing the vertical pressure, keeping the abrasive grains in position and reducing the transverse pressure. Minimum frictional heat is generated, and no abrasive grains are forced into the work surface. The finish itself is optically smooth and free from fragmented or smear metal, and the surface after superfinishing is a true, geometrically-developed, wear-proof bearing-area, free from oil-bearing, rupturing protuberances and accurate to within sub-microscopic range.

The abrasive is silicon carbide or aluminium oxide in block form and with a grain size of 180–600 according to the fineness of surface required—600 is advised for steel. A vitrified bond is used for the harder steels, and a shellac bond for low carbon steels. Table XIII shows the number of stones as governed by the work diameter. They may be rectangular, square, curved, dish-shaped or of other suitable form, and are 4–6 in. long. Motion is random, and pressure is about

SURFACE BROACHING

20 lb./sq. in. of the stone area. The lower the pressure, the better the finish. The normal range is from 1·3 lb./sq. in. for pistons, to 18·7 lb./sq. in. unit load for punching brake drums. Variable pressures are often used at the start, and for greater areas higher pressures of about 30 lb./sq. in. may be used.

Paraffin or a light thread-grinding fluid together with a blended mineral oil of fairly viscous type to the extent of 10–20 per cent is the most effective. The heavier the lubricant, the higher the pressure and the slower the cutting. Lubricant must be thoroughly filtered, but delivery pressure is not high. Traversing feed ranges from $\frac{1}{16}$–$\frac{3}{8}$ in./rev. according to the work diameter, the best range being $\frac{1}{16}$–$\frac{1}{8}$ in. Traversing is not always necessary.

The superfinishing machine is made up of a headstock and tailstock with centres or a collet. The head holding the stones rests on two horizontal bars along which it travels. Power is provided electrically for both traversing and oscillatory motions. A good standard machine takes cylindrical work up to 6 in. dia. × 18 in. long, but also deals with flat surfaces. The degree of surface finish given is from 0·002–0·0001 in. according to the operation.

TABLE XIII

Work Dia. (in.)	No. of Stones
Up to 3	1
3–7	2
7 and over	2 or 4

Surface Broaching See **Broaching**.
Surface Gauge See **Gauges**. A measuring gauge to enable the difference between the height of some part of the work and that of the work-table of a machine tool, such as a planer (q.v.), to be ascertained, after which it is transferable to some other point.
Surface Grinding Machines For grinding flat surfaces of work held in place by magnetic chucks, **horizontal spindle** machines are mostly employed, of three classes, namely **reciprocating table** machines, **rotary table** machines and **face grinding** machines. The first class are used in various sizes according to the operating area of the work-table, and are the most popular. The second have inclinable or fixed work-tables, the inclined-table machines grinding mostly concave or convex surfaces, but being suitable, when required, for flat surfaces. The third are also much used, and are

larger than the other two classes, the work being carried by a vertical table, either movable or stationary; in the latter case the grinding wheels move over the work, and are of either cylindrical or segmental type. (See **Grinding**.)

For mass-production power-operated feed and traverse machines are the most suitable, but hand-feed machines may be better for tool-room work. Special machines are manufactured for particular operations.

Some surface grinding is carried out with vertical-spindle machines using either reciprocating or rotary work-tables, and these are particularly suitable for work with two opposing flat surfaces. Large or small quantities can be dealt with, and some machines are exceptionally good for producing parts having great accuracy to size, parallelism and flatness. They will accept large table-loads and are mostly applied to parts requiring maximum limits of 0·001 in. and surface finish of not more than 32 micro-in.

Surface grinding is also done on double-wheel machines for parts having opposed flat surfaces of identical form that give each other support, and these are uniformly ground as they go through their cycle. Double-wheel machines are economical when a high rate of output is required, but give fine limits on thickness and flatness. See **Grinding**.

Surface grinding is also achieved by **electrochemical grinding** (q.v.), using the largest possible wheel, of a width capable of covering in a single pass the entire area to be ground. The operation is best carried out with a power-feed, which must be carefully controlled. Another technique is **electrochemical discharge grinding** (q.v.) using a wheel of graphite, which costs less than an abrasive wheel. This process is suitable for delicate work difficult to grind by alternative methods.

Swarf The metallic dust produced by grinding-wheels, mixed with particles of abrasive or other matter, and water.

Swing In a lathe, the dimensions of the component or piece that can be cut in it. These dimensions are normally those of the gap, over the bed, slide-rest or carriage, and between the centres.

Swing Table In a drilling machine, the work-table swings about the central pillar or frame, as the construction directs, thus enabling any required area of the component to be located beneath the drilling tool.

Swiss-type Bar Machine (See **Multiple-operation Machining**.) A single-spindle, automatic bar machine having 5 radial tool slides in the same plane, used for generating radii.

T

Tabbing The method of reclaiming blanked-out parts after etching when preparing masters for **chemical machining** (q.v.). Small lines are added to the artwork to produce "tabs" or connections holding the parts together after etching, and are removed from the work later.

Tailstock Sometimes termed **loose headstock,** this is situated on the right-hand side of the lathe operator, and its centre is carried in a barrel which is internally screwed at the opposite end from the cone (see Fig. 55). Within this barrel is a screw operated by a hand wheel which enables the barrel to be so moved that the cone can be adjusted when the tailstock is fixed. This screw also has a cone-shaped centre. The centres carry a cylindrical object and hold it firmly, but not so firmly as to prevent it from rotating; however, the work must be so held on the centres that its rotation is about a true axis.

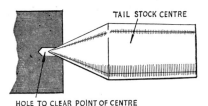

Fig. 55   Lathe tailstock

Many cylindrical parts turned between lathe centres are drilled with central tapering holes to accommodate these centres. In some modern lathes revolving centres are provided in the tailstock, which can be moved along the lathe bed longitudinally to suit the length of the work or to free it from the operating position when out of use. The spindle carrying the supporting centre is adjustable to suit minor variations in length.

Tallow An animal fat melted to provide a lubricant in some machining operations or a buffing material for **honing**. It is expensive, but for honing is the most economical material owing to its long-service duration. It usually flows under pressure, adheres to metallic surfaces more satisfactorily than mineral oils, and is often an ingredient in proprietary oils for **buffing**.

Tandem Planing A method of planing, sometimes known as **Gang Planing,** in which rapid loading and unloading of the work are achieved by special fixtures, with a considerable rise in output. The parts are either continuously planed by being firmly held in place with ends or surfaces abutting, or are held 6–8 in. apart. Any distance less than this causes fracture of the tools owing to chatter as they leave a cut and pass to the next. See **Planing**.

Tang The stem or shank of a cutting tool inserted into a socket, as with drills, or a handle, as with files. This part of the shank is not usually hardened.

Tangential Chasers In self-opening dies for die threading, groups of chasers having cutting-edges that lie along a single face of the die. A bevel or chamfer portion embodying the first complete tooth produces the first thread contour, and in addition takes off surplus metal from any work above the required dimensions. This bevelled portion lasts throughout the service-life of the chaser and is not re-sharpened. The angle of rake depends on the properties of the material cut and is varied as required. These tools, owing to their long service-life, are specially advantageous for making threads in metals of some hardness. They can be resharpened many times, but this resharpening is not carried on after the chaser length has become too short for the die to hold it or for it to be gripped for sharpening. (See **Die Threading.**)

Tangential Rolling A form of **thread rolling** (q.v.) in which the rolls are carried past the work in a line parallel to the radial. This brings the roll-axis opposite to the work-axis and the pitch-line of the contour of the thread tangential to the work surface. The forward movement of the rolls brings them to a point at which their centre-line directly opposes the work centre-line, and at this point the deepest penetration occurs. Its depth is governed by the degree of offset of rolls in relation to work, and only a small amount of axial motion takes place between rolls and work.

The operation is carried out in lathes or automatic bar machines embodying single- or double-roll attachments located on the cross-slide, and the rotation of the rolls is achieved by their frictional contact with the rotating work. The more popular attachment is the two-roll. The machines can be had in different dimensions, each of which accepts a range of work up to $2\frac{1}{2}$ in. diameter, but machines for larger diameters can be obtained when necessary.

The attachments allow of adjustment and regulation to suit particular work dimensions, these operations being performed in the attachment and not by the motion of the cross-slide. **Two-roll tangential attachments** give heavier roll-pressures than when **radial in-feed rolling** is used, and in addition radial motion between the roll-spindles is absent during the operation. The threads have to withstand bending-loads a little heavier than with a corresponding two-roll attachment of radial-feeding type, but the threads can nevertheless be rolled on comparatively hard metals fairly widely separated from the collet. Changes of speed during tangential rolling are not required.

The drawbacks of the two-roll attachment are the dimensions and capacity of the equipment. High lead angles and deep blunt forms should be avoided owing to axial movement. On the other hand, the single-roll attachment is liable to bend the work and is restricted to short threads rolled close to the collet in soft metals, such as low carbon steel or non-ferrous alloys, not exceeding in hardness Brinell 197. Threads are not usually longer than $\frac{7}{8}$ in. with this attachment. (See **Threading**.)

Tangential Tools Tools cutting in a tangential direction, the edge and face passing beyond the axis of the work to develop the desired contour. While these tools are less popular than they originally were, some users prefer them for finish-turning of brass and aluminium or its alloys. Often, however, there is nothing to choose between these and **radial tools** if the conditions allow either to be used. (See **Form Tools**.)

Tantalum Carbide A carbide produced by powder metallurgy and used in conjunction with **titanium carbide** (q.v.) as an ingredient in cemented tungsten carbides, the two carbides together being present to the extent of 2–5 per cent. It is also used, however, as a cutting-alloy in its own right, a typical composition being 5·5–16 cobalt, 18–30 tantalum carbide per cent, which alloy has a density of 14·8–13·5, and is used primarily for heat- and shock-resistance rather than for cutting-tools. There are, however, some cutting-tools with a mainly tantalum carbide ingredient, the titanium carbide being present to a less extent, the two carbides together totalling from 12–25 per cent according to the purpose. These are applied primarily to cutting steel where resistance to wear and to formation of craters on the tool is required. The amount of carbides present determines whether the material is for light or heavy cuts. Tools containing titanium carbide and a lesser amount of tungsten carbide are not as expensive as these, and are therefore chosen when the economy of the cutting operation makes this factor important.

Special grades are made for use on steel, and only very small tools are made solid, the rest being tipped tools with tips brazed into position, then finish-ground and if necessary lapped and honed to give a final keen cutting-edge. The shank is of mild steel or a specially toughened carbon steel.

Tap Grooving Milling-out in a suitable machine the flutes or grooves of **taps** (q.v.), so producing their cutting-edges.

Tape-controlled Machine Tools Machines such as lathes, boring machines, drill-presses and milling machines, operated by numerical control (q.v.).

Taper Attachment See **Taper Machining**.

Taper Chamfer Taps In **tapping** (q.v.) a form of solid taps designed to spread the cutting-load over the maximum number of threads so that the tap may enter the work without difficulty. They are particularly advantageous in work on the more difficult materials, though not recommended for cutting blind hole threads. Furthermore, full threads in open-ended holes need a greater length of travel than they would otherwise.

Taper Machining Machine tools such as lathes sometimes have their capacities heightened by an attachment, such as a taper-attachment, altered to suit different cutting-angles for the turning or boring of work needing a tapered form. Such attachments, carried by the cross-slide, enable the lathe-centres to be maintained in alignment so that the work is carried out with greater accuracy. When single-point cutting-tools are used, the variation of taper-diameter is governed by the rotational axis of the work in relation to the lengthwise motion of tool and carriage in both vertical and horizontal planes. Unless completely parallel relation is achieved, the work tapers.

Die heads can be used for taper turning in multi-operation machine tools. Taper can also be produced by **chemical contour machining,** slow immersion or withdrawal being necessary to ensure the maximum degree. Step-tapering removes less un-needed weight than continuous-taper machining, but is less expensive and enables design to be more flexible.

Taper Reamers A form of **reamer** (q.v.) producing either the type of tapering hole suitable for holding tools with taper-shanks, such as twist drills, reamers, etc., or taper pins.

Taper-shank Oil-hole Drills See **Oil-hole Twist Drills.** As compared to drills of this type having straight shanks, the hole is bored with greater concentricity, speeds and feeds can be increased and the drill is less susceptible to bending-stresses.

Taper Threading Cutting threads in a tapering pipe calls for stricter regulation and greater power than does the cutting of straight threads, since a larger number of teeth are cutting simultaneously. The usual machines are employed (see **Threading**) and a positive lead-control is adopted. The tools used may be solid, non-adjustable or adjustable, the latter being more suitable when a large amount of metal has to be removed, and when strict regulation of pitch-line taper is desired.

Collapsible taps allow the work to be automatically withdrawn after the operation, with no reversal of rotation. They allow of shorter time-cycles, and lengthen tool service-life, but are more expensive and more liable to clogging with imperfectly filtered or non-filtered cutting-fluids. They also require more upkeep than the less complex tools. The same cutting-fluids are used as in ordinary machine threading, but must be filtered and continuously supplied.

Taper-thread Taps Taps designed for producing threads having a uniform reduction in pitch diameter as between thread and thread. See **Tapping**.

Tapping The machining operation whereby a screw thread is formed in a hole by tools known as taps, the tapping-diameter of the hole corresponding to the core diameter of the thread. The tap is a cylindrical or cone-shaped cutting-tool possessing threads of suitable contour on its circumference.

The hole is first drilled in the work, and has a rather larger diameter than the nominal minor screw diameter. The tap is then passed through the hole, rotating and being kept absolutely straight and in line with the hole-axis. Sometimes a jig is used for this purpose. The drill-press is used for solid taps if no other type of machining work has to be carried out, and has a tapping-head whereby the spindle is swiftly and accurately arrested to maintain depth limits, and is easily reversed for the clearing of chips.

Such heads are not necessary if collapsible taps are used. The tap passes down the hole to a specific point, at which it automatically collapses and withdraws from the hole so that the spindle can retract without pause or reverse. If lead-control is necessary it is better to employ a tapping machine proper.

The machines used are either general for standard work, or special for use on a single class of work. They may be vertical or horizontal, single-spindle or multiple-spindle. Drill presses, manual or automatic turret-lathes and multiple-operation machine tools may all be used. In some machines the spindles are controlled to modify the distance between centres, or fixed for repetition work, such as the tapping of nuts.

The choice of machine depends on the dimensions of the work, the form of the piece, the output desired, the dimensional limits, the degree of surface-finish required, the number of related operations, and the economic considerations.

TAPPING MACHINES

There are seven classes of tapping tools used (see **Taps**), namely solid, shell, sectional, expansion, inserted chaser, adjustable and collapsible. These are made of high speed steel for the most part. Some use is made of carbide chasers (see **Tungsten Carbide**). The tools may be given some kind of surface treatment, such as nitriding, chromium-plating or steam-oxidation; to lengthen their service-life and give the hole a better surface; to minimize chip build-up on the work metal; and to lessen erosive wear by chips. However, these result are not always achieved, and in any event the treatments are expensive and cumbrous. Tapping-tools are largely chosen on the type of material to be cut, the type of thread desired and the expense.

The cutting-fluid used is most important as tap teeth are extremely sensitive to heat, and the crowding of chips in tapping creates excessive heat. The fluid is supplied under pressure and filtered to ensure minimum abrasion by particles of swarf. Sulphurized or chlorinated oils, soluble-oil emulsions and mineral oils are used according to the type of work and material.

Tapping Machines Machines used to form threads in holes, and to carry out a wide range of tapping work. They may be either single- or multiple-spindle machines, and have an automatic reverse so that the main spindle can be withdrawn by reversal of the spindle motion. The machines are not difficult to control and make holes in nuts, crankshafts, studs, oil-pans, valve-covers, oil-pressure plungers, etc. They may be either vertical or horizontal. Some of these machines will tap a series of holes at one and the same time. Some have spindles adjustable to modify the distance between centres, others have fixed spindles and are primarily used for tapping duplicate components.

Some have a variable mechanism to give the forward or reverse motions of the tap-spindle and regulate these movements. Some do other work, such as drilling holes, if necessary. In a typical machine for tapping nuts, the work is fed into a hopper, and a series of blanks situated on the tap shank guides the tap in its holding device. The nuts are moved into position for threading and travel along the tap shank until ejected. Two spindles work separately from each other, enabling two different classes of work to be tapped together, each occupying one of the spindles, and to shift from one size to another presents no difficulty.

The tap-reversal speed is double that of cutting, for economy in time. As soon as the tap passes into the hole in the nut the thread-lead automatically carries on the feed, and when the work is finished, the tap is reversed and withdrawn.

TAPS

Taps Screwed plugs having precisely-formed threads, accurate to dimensions, given cutting-edges running longitudinally, and screwed into a hole manually or mechanically to produce an internal thread. They are of seven main types, the first being **solid**, i.e. made in one piece, mostly of high speed steel, though some are of carbon steel or tungsten carbide. They have either straight or taper threads, the first for producing unvarying pitch-diameter, the second for those with a uniform reduction in pitch-diameter. They have flutes and a degree of bevel or chamfer. (See **Chamfer.**)

Shell and **sectional** taps are of high speed steel, have no shanks, and their thread covers all or most of their length. They are mounted and driven by an arbour. The sectional taps have inserted shanks and are not so long. **Expansion** taps have an axial hole running from the front end to beyond the threaded portion of the tap, as well as a slot or slots running radially from the tap-surface between the flutes and covering the entire operating length of the threaded area to give it a degree of spring. These are used primarily for finishing or for forming screw-threads in free-cutting steels. They are somewhat expensive.

Inserted chaser taps have slots to receive a number of chasing tools, secured by various devices. They are high in first-cost, but are economical for long runs, and can be used for tapping holes oversize.

Adjustable taps having ground thread chasers with serrations on their backs are for finishing to exceptional accuracy, but have to be specially ordered. **Collapsible taps** withdraw in a radial direction on completion of the cut, without reverse rotation. They may be stationary or revolving, and have flat or blade-type chasers, though for large diameter work circular chasers are often employed. They are readily adjusted, suitable for short or long runs, and minimize frictional wear, risk of tap-fracture, or injury to the threads caused by chips. On the other hand they necessitate a more robust machine, need more upkeep, are not suitable for holes less than $1\frac{1}{4}$ in. dia., and are high in first cost.

When taps are power-tapping, using an ordinary drill-press, special **tap-chucks** reverse the tap-rotation on completion of cutting. In one type of tap-holding chuck the tap is automatically arrested as it makes contact with the bottom of the hole, or an adjustable depth-gauge hits the top of the work. Elevation of the spindle then reverses the tap and withdraws it at a greater speed.

Template (sometimes written **Templet**). A thin sheet cut to the form or contour desired on a finished surface and employed as a model or gauge in machining. In making form-tools, for example, it may be applied to both roughing and finishing operations, but in

THREAD CHASERS

most modern workshops this method is no longer used. In machining bevel gears, however, the template is used for low production, nongenerating, with the least possible tooling. The template is several times as large as the tooth to be machined, and high accuracy is possible. Two templates are used, one for each side of the tooth, and covering a small range of ratios. 25 pairs of templates cover all 90 deg. shaft angle ratios from 1:1 to 8:1, for either $14\frac{1}{2}$ or 20 deg. pressure angles, the technique being founded on equal-addendum tooth-proportions for all ratios.

Contour planing is sometimes carried out with a template, as in the cutting of helical grooves in large diameter rolls. **Shaping** also uses templates for automatic duplicating, the templates, cut out of sheet steel, being set in a suitable holder and the blanks contour-shaped by high speed steel tools. This is claimed to be the most economical method, a new template being made whenever a shape-cut needs to be modified.

Thread Chasers See **Threading Machines.**

Thread Gauge A gauge having a screw thread.

Thread Grinding The forming of threads on a part by grinding rather than machine-cutting or **thread-rolling** (q.v.). The threads are produced by rotating the work and bringing it into contact with a revolving abrasive-wheel formed to produce the desired thread. Wheel and work move axially to one another conformably to the pitch of thread, and the operation may be either internal or external. The operation corrects small inaccuracies of shape or dimensions set up by heat-treatment or previous machining.

Both cylindrical and centreless grinding methods are used (see **Grinding**), the latter giving a higher productivity and being mostly preferred. The machines may be external, internal or universal. In general they provide a precise axial movement of the wheel in relation to the work, mostly by means of the lead-screw, to generate the thread. A means of truing or dressing the wheels is embodied, and also mechanism for inclining the wheel to cut the necessary helix. The wheel may have one rib or several and has to be adequately supported. The abrasive used in the wheel is governed by the hardness and character of the metal to be ground and by the number of threads/inch. Mostly aluminium oxide wheels are used, with silicon carbide for titanium and diamond for carbides and ceramics. Grit-size is controlled mainly by thread-pitch and grade by grit-size and the material of the work. The finest threads require the finest grit, and the fineness will also determine the surface-finish and wheel-hardness. Vitrified or resinoid bond is used.

Coarse grit takes off the metal faster, as does a resinoid bond,

THREAD MILLING

which is the best for quantity production, but vitrified wheels are better for accurate grinding and rectifying errors in already-existing threads. A mineral-based sulpho-chlorinated oil is mainly used as coolant, and is only abandoned if for some reason it proves inadequate, in which instance a specially-blended proprietary oil may be adopted.

The wheels will grind cylindrically, and in other operations, e.g. centreless grinding and cylindrical grinding, with multi-ribbed or single-edged wheels or those with a single rib. The multi-ribbed wheels are for high production if the form of the work is suitable.

Cast iron, nickel alloys, stainless steel and tool steels may all be ground with screw-threads by these methods.

Thread Milling The cutting of screw-threads in a thread-milling machine with a **milling cutter** (q.v.), using standard or planetary thread milling machines of either **universal** (q.v.) or **production** type, when high accuracy together with a better finish is required; where the pitch of the threads is too coarse for die threading; where the thread is close to a shoulder or other obstacle; and where the part is too big or complex in form for normal threading. The machines, semi-automatic, embody a master-screw and stationary segment rather than a lead-screw. These have to be changed whenever pitch or lead changes, a no-lead attachment producing annular grooves. The speeds range from 45–140 surface ft./min. according to the material, the lowest being for precipitation-hardening stainless steel, the highest for low carbon steel of free-cutting type.

The cutters are of either single- or multiple-thread type, the former used for coarse pitches and threads longer than a multiple-cutter will handle, the multiple-thread cutters for cutting on the sides and root of the thread only.

Thread-rolling Producing threads on components by passing them between rolls under pressure. The work rotates and receives the required impressions or annular grooves from hard steel dies or rolls to which the desired pitch and form of screw-thread has been given. No material is machined off the work, the surface metal being forced upwards to create a thread having roots and crests. In addition to rolls, flat dies are used, these being traversed across the face of the work. Cylindrical or roll dies of radial in-feed, tangential-feed, through-feed, planetary and internal types are used. The operation is for most materials except grey cast iron and inductile metals such as the carbides. Any required standard thread form can be produced by rolling, and the process also forms splines, helical and annular grooves, knurled surfaces and involute gear teeth, as well as burnishing.

Thread-rolling gives a harder and stronger metal with high fatigue strength (as compared to material not stress-relieved) if stress relief by heat-treatment is given to the metal after thread-rolling. Smooth surface of threads and roots free from blemishes minimize wear and fatigue of the part threaded by rolling. Thread-rolling is probably the method best suited to high production and large quantities, but these are not the only factors to be considered in deciding the operation to be used. For example, the smaller the number of pieces, the less economical thread-rolling becomes as compared to thread-cutting. It would be uneconomical to thread-roll less than 7,500 screws 6 in. long by 2 in. long threaded section. Aluminium alloys, brass, copper alloys, heat-resisting steels, Monel metal, stainless and other steels, can all be thread-rolled.

Threaded Adaptor A threaded plug inserted at the headstock end of an engine lathe with a male centre in the tailstock, enabling 7 operations on tubular parts to be carried out in a single chucking.

Threaded Grinding Wheels In grinding gear wheels, form-grinding with a wheel having a suitable thread. The operation is confined to spur and helical gears up to 20 in. dia. and having a diametral pitch of 24 or coarser. The threaded wheel is claimed to give greater accuracy and speed and easier modification of tooth-form than any other method. However, the operator requires a high degree of skill and the machine used is intricate and sensitive.

Threading-dies These are **dies** (q.v.) used for cutting threads, either internal or external, on such parts as small and average-sized screws, bolts, studs and heavy screws up to 4–5 in. dia., or above in special instances. The dies complete the operation in a single cut. They may be solid or self-opening. **Solid dies** are used in multiple-spindle machines and those that do not take a self-opening die, and some are also used in hand-operated machines. The multiple-spindle solid-die machines give a much higher output, but the hand-operated machines give only small quantities. **Self-opening dies** (q.v.) are more expensive to buy, but their renewal cost is lower, and on the whole they are more economical. They are easily taken out and minimize thread-damage by crowding chips. Their chasers have a longer service-life.

The threads are actually cut by **chasers** (q.v.) used either individually or in sets. **Single chasers** are for when the work rotates between centres and the die is unable to enclose the component, or where there is insufficient room for the die using sets of chasing-tools. Single-chaser dies thread much quicker than single-point cutting-tools, but are not much used for high production. **Self-opening dies** are more readily adjusted, thread with accuracy up to a certain

THREADING MACHINES

limit, give better chip-clearance and a better finish, and are quickly removed from the work, which greatly faciliates production. They do not produce **stop lines** (q.v.). (See **Threading Machines**.)

Threading Machines Threading can be done in a variety of machines, such as the engine-lathe using a single-point cutting-tool, which produces the threads by turning. Lead-control can be provided by the lathe. Drill-presses are readily set up and easy to work, and are often used for threading. Hand-operated turret-lathes are used when the quantities are comparatively small and threading is combined with other operations. Only if these machining operations are combined does the hand turret-lathe become economical because its rate of output is lower than that of a drilling machine, but it is useful for work too large for the drill-press. Lead-control is sometimes introduced to simplify the work for the operator.

Automatic machine tools such as turret-lathes, single- or multiple-spindle automatic bar or chucking-machines, etc., are used only when a cycle of operations gives greater productivity. Special die-threading machines with lead-control embodied in their construction are used for cylindrical threading or for work of complicated form. Some of these deal with rod, shafts and pipes requiring threads.

Milling machines are sometimes used, but are confined to those classes of work requiring exceptional accuracy or where the form and dimensions of the part make alternative methods unsuitable. However, the operation is costly. (See also **Thread Grinding, Thread-rolling, Threading,** etc.)

Three-co-ordinate Optical Measuring Machine A measuring machine with all settings on projection screens, reading to 0·00005 in. (0·001 mm.). There are also universal machines of this type capable of high precision; and even larger machines with optical settings on projection screens and electronic micro-indicators.

Three-ribbed Grinding Wheels In grinding threads (see **Thread Grinding**) a wheel having three ribs on the edge, the first rib removing about $\frac{2}{3}$ of the metal, the second removing what is left, apart from about 0·005 in., which is surface-finished or skimmed off by the third rib. Sometimes a final grinding is done by a fourth rib whose top has been ground flat to finish off the thread crest.

These wheels give accuracy higher than, or at least equal to, that achieved by **single-ribbed wheels** (q.v.). They can be sloped to a specific helix angle with a proper allowance for the work-curvature radius, and they traverse the face of the work instead of cutting deeply into it and then withdrawing for the next cut.

Through-feed Centreless Grinding In centreless grinding, a technique in which the work travels either sideways or axially between the abrasive and the controlling wheels. This method is used only when the work is plain and cylindrical and has no steps or obstacles. (See **Centreless Grinding, Grinding,** etc.)

Through-feed Rolling In the rolling of screw-threads (see **Thread-rolling**) the method of passing the work axially between rolls or dies that have a lead-angle differing from that of the part to be threaded, thus enabling it to feed. The thread is progressively formed by giving the dies an initial taper at their entering end. Taper, of small amount, is also given to the other end of the dies to ensure that the rolling pressure is not taken off too quickly, so producing stop-lines on the work.

Cylindrical die thread-rolling machines mostly embody end-feeding attachments, the dies being relieved at each end. There is no restriction on the form of thread produced by this method, which is particularly useful for threads of extremely blunt form, being superior in this to **In-feed Rolling** (q.v.).

Spur and helical gear teeth can also be formed by through-feed rolling.

Tipped Tools Cutting-tools made up of a body or shank and a tip of some hard cutting alloy, such as high speed steel, tungsten or other carbide, stellite or a ceramic material, enabling expensive materials to be economically used for cutting difficult materials, whereas in solid tool form they would cost far too much.

Titanium Carbide Tools Tools made by sintering titanium carbide and using it in combination with other carbides, such as **tantalum carbide** (q.v.) or **tungsten carbide** (q.v.) in the form of cutting tips of tools used in machining hard materials. Some tungsten carbide tools contain mostly titanium carbide as an addition; others have only a small amount of titanium carbide added. The first group are used for either light finishing-cuts on steel at high speed, medium-cuts on steel at medium speed, or roughing-cuts on steel. The second group are for light cuts on steel where durability of the cutting-edge and resistance to cratering are required, and for general cutting and heavy cutting of steel where abrasive wear from the rough scaled surface of castings is likely, and durability of cutting edges and cratering resistance are required.

In general, the carbides of the first group have only small shock-resistance, and are less dense than purely tungsten carbide tools, so that they weigh less. Titanium-carbide-containing cutting-tools are less expensive than tantalum-carbide-containing tools.

Titanium carbide has been used in the form of insets for improving the surface-finish of steel forgings to be burnished after machining. The work is done in a multiple-operation machine tool using a

TOOL ANGLES

single-spindle chucker and triangular indexable inserts. Titanium cutting-tools have also been used to advantage in turning, facing and chamfering cluster-gear blanks in an automatic lathe, a higher surface finish, a higher speed and a lower cycle time being achieved than with standard steel-cutting tungsten carbide tools.

Titanium carbide is the principal ingredient in **cermets** (q.v.), being present to the extent of 70–80 per cent. These are for light finishing cuts, being more brittle than the ordinary carbide tools.

Tool Angles The cutting angles of **single-point cutting tools** (q.v.), etc., are of great importance. In Fig. 56 a section of a tool perpendicular to AB is shown. The three angles α, β, γ define the properties of the cutting-edge. In Fig. 57 if GH is the surface being cut, then α is the **clearance angle** of the tool, and γ the **rake angle**.

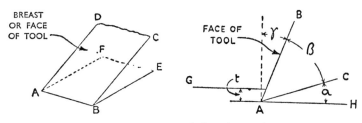

Figs 56 and 57   Tool cutting angles

Many tools do not have a straight edge, but irrespective of the form or type of tool, wherever a section can be taken perpendicular to the cutting-edge at this point, these three angles, α, β, γ define the characteristics. Of these α and β are the most important, but the strength of the tool depends in some degree on α and β. The sum of the three angles is always 90 deg.

Cutting-tools as shown in Fig. 58 are often ground with the face sloping away from the cutting edge. The angle γ^1 is termed the top- or front-rake, but this is not the true rake angle. A section at E–F shows the side-rake, but again the angle G–E–F is not the true rake angle. The difference between the true rake angle and the commonly-accepted terms, front- and side-rake, should be clearly recognized, because the performance of the tool depends on its true cutting angles, α, β, γ, irrespective of tool form, and these for any specific material should as far as possible be the same.

As generally defined, the two types of rake angle are **top-** or **front-rake** (the angle measured along the tool length, i.e. towards the butt) and **side-rake** (the angle measured at right angles to the

TOOL ANGLES

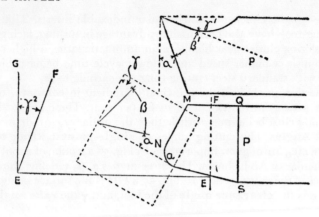

Fig. 58   Cutting-tool rake angles

tool or the downward slope from the cutting-edge). As explained, neither is the true rake-angle. The true rake-angle depends on the material to be cut, and facilitates chip severance. Tools used with **negative rake,** i.e. with a rake angle of 0 deg., or "without rake", as is commonly said, are used in threading because the thread has to be given the same form as the tool (see **Threading**). A 0 deg. angle is also used on occasion for cutting some non-ferrous metals and chilled iron. Recently **negative rake cutting** (q.v.), especially with milling cutters, has given excellent results.

The size of the **rake angle** varies with the material and the machining operation, being smaller as the work is harder to give the cutting-edge maximum strength and prevent it from spalling under the pressure of the cut. Easily machinable steels demand a larger rake angle or the chip-pressure on the tool face becomes more difficult, while cutting speed has to be reduced.

The **side-rake angle** is as far as possible that at which rake and clearance angles are the same at all points of the cutting-edges. The **wedge angle** is governed by clearance and rake angles, and is larger for hard materials. Maximum cutting speed on steels when roughing is attained with a wedge angle of 61 deg. At 50 deg. angle the tool-pressure is less, but the cutting-edge is slighter and crumbles more easily. On the other hand a 90 deg. wedge angle gives a robust edge, but the chip curls up and spirals, causing severe friction and wear on the tool nose and impeding the cutting edge as it attempts to penetrate the steel. The **clearance angle** need be no greater than suffices to ensure easy, unhampered cutting. For turning steel in the lathe, an angle ranging from 5 deg. is satisfactory. Larger clearance angles are usually needed for non-ferrous alloys. Tungsten carbide tools have mostly a smaller clearance angle of 4–6 deg. The clearance angle should never be too great or it will weaken the cutting-edge and cause it to snip.

Considerable modification of the effective rake angle and clearance angles is made by elevating the tool above the work, or lowering it below, i.e. above or below the horizontal centre.

Some tools such as **parting-off tools** (q.v.) and right-angle **recessing tools** (q.v.) need clearance of about 10 deg. on the tool front or flank, but clearance is also given to the sides, usually about 3 deg. The side clearance angle is similar to the front clearance angle. The rake angle is made small or even 0 deg. because these tools are liable to dig-in. Parting tools are sometimes used for cutting grooves in face plate or chuck work, in which case one side is given a larger clearance angle than the other.

Tool-changing Station A rotatable tool-storage magazine consisting of a series of turrets that take and store the tools. The turrets vary in size so that a wide range of tools is handled. (Brit. Pat. 1,146,253.)

Tool-rest See **Slide-rest**.

Tool Shapes Fig. 53 shows modern standardized tool-forms covering a wide range of machining operations in lathe, boring, shaping and other machines. These shapes are adapted for butt-welding, the line across tool neck or shank indicating where the top rake runs out, but the shapes and sizes do not exhaust the number of forms in use, though many eccentric shapes still in use have no genuine justification. The tools shown in Fig. 53 produce most of the work of modern machine shops. See pages 182–3.

There are, of course, special **form tools** (q.v.) whose cutting-edges produce an unusual contour, or which have the tool-nose bent in a peculiar way to present the cutting-edge to some otherwise inaccessible area.

Toolholder Bits Known in America as **tool bits** these are small, solid pieces of high speed steel ground by the user to a desired tool form, and inserted in a special toolholder. They are hardened before being ground, and when ground are ready for use. Regrinding is easily carried out. They use relatively little high speed steel, and are easy both to remove and to adjust to the correct height. Some toolholders hold the bit with its axis at an angle corresponding to the desired rake angle, whereas others locate the axis to give the best clearance angle. With 0 deg. rake the bit is held in the horizontal position. It is secured in the holder by two set-screws or sometimes one only, and set to give the requisite side- and front-clearance. The holders may be either right- or left-hand. Any class of high speed steel commonly in use can be supplied in the form of toolholder bits.

Toolpost A circular or cylindrical post having a screw at one end and a circular flange at the other, secured to the upper part of the slide-rest of a lathe to hold the cutting-tools firmly down. The flange fits into the bottom of a tee block in the slot of the tool-rest. These posts are employed only in the smaller sizes of lathe, the larger machines using various types of recessed studs, clamps, etc. The post not only provides support for the tool, but also holds it tightly in position. In some types of post the height of the tool can be changed and the tool locked in place, without any modification of cutting angles.

Top-rake Angle See **Tool Angles**.

Torque The turning-moment produced by a tangential force acting at a distance from the rotational axis or twist, a factor taken into account when twist drills, taps and other tools are liable to encounter a torque-producing force. In twist drills the average torque for $\frac{13}{16}$, $\frac{7}{8}$ and 1 in. is 40–50, 45–55 and 50–60 ft./lb. respectively. Torque is often overcome in these tools by butt-welding, the cutting-portion carrying the helical flutes being butt-welded to an oil-toughened carbon steel or alloy steel shank, and less likely than solid high speed steel to fracture when subjected to these forces.

The amount of torque taps have to withstand is governed by the work and its hardness, the form of the tap, the surface-speed, the extent to which a full thread has to be cut, how the tap is ground and the coolant used. The tougher and stringier chips given by some metals increase the torque-thrust. Cast iron has minimum torque when tapped. In tapping steel straight-fluted, **plug-chamfer taps** (q.v.) have a higher torque to withstand than the spiral-pointed type. In tapping some materials an 11 deg. hook angle may reduce torque when cutting low carbon steel. The influence of speed on torque is slight. Full threads produce higher torque when tapped owing to the greater amount of metal taken off. Grinding produces lower torque in all circumstances if an automatic chamfer-grinding machine is used, the manner in which the chamfer is ground producing considerable variation in service-life of the tap. Careful choice of coolant reduces torque. Faulty alignment of the tap owing to an excessively rigid holder increases torque.

Tracer-control A method of machining in which a stylus or cutting point follows the form of a master-copy or **template** (q.v.) of the part to be machined, and passes on information so gathered to the controlling elements of the machine. In this way the form and dimensions of the master copy are transferred to the work to be machined. The lathe can then produce exact copies of the master-part far faster than with hand-operated lathes. Tracer-control alone is unable to produce parts automatically from information supplied, but is less expensive than specially-built centre-lathes embodying hydraulic-copying devices.

A fully-equipped tracer-lathe embodies one or more cross-slides so that it may rough- or finish-face, back chamfer, groove or undercut. If required, automatic-positioning steady-rests and an automatic-indexing tool-head to accommodate roughing and finishing tools can be incorporated.

Tracer-control is also used in boring, planing, milling and shaping machines.

Transverse Shaving See **Shaving.**

Traverse The extent to which a tool travels lengthwise along a machine tool bed, the motion being provided by either a leadscrew or a rack and pinion. The term is likewise used for various other machine-tool motions.

Trepanning Machining a circular hole or groove in solid metal with a tool in which a cutter or cutters rotate about a centre. The cutters are mostly **single-point tools** (q.v.). Alternatively, the tools cut out a disc, tube or cylinder, the central portion of the material being removed as a solid piece. The tools cut narrow grooves.

Trepanning gives accurate results also when producing deep holes in the bores of gun barrels. In general the method is for small numbers of parts when blanking-out with dies would be uneconomical and where such special operations as flame-cutting would be impracticable owing to the absence of the necessary apparatus.

The technique will produce solid round discs from flat material; large shallow holes running right through; circular grooves; and deep holes. For round discs an adjustable fly-cutter mounted on a twist drill, which acts as driving-member and pilot, can be used. This tool necessitates a centre hole, but if this cannot be allowed, the tool may be used without a pilot, but with more chatter and less control of dimensions. The maximum size of disc cut is 6 in., and the maximum thickness about $\frac{3}{4}$ in. Speed is not more than 35 surface ft./min., and feed hand-controlled. Cutting-fluid is not required.

Large shallow through holes are often cut with single-point cutting tools of tungsten carbide having inserts in the side brazed into position. Trepanning often produces these holes with greater dimensional precision and efficiency than drilling, but the governing requirement is that the diameter of the hole to be cut should be about 5 × depth or more.

Circular grooves are machined with a suitable cutting-tool that produces the required form of cross-section. Often the best results are obtained with a combined drilling and trepanning tool on the lines of a hollow mill. This consists of a twist drill held in a hollow cutter by a set screw, with two or more cutting edges and a back rake angle of 20 deg. Drilling machines are used to drive the tool.

Deep holes are cut with tools similar to those used in **gun-drilling** (q.v.). Trepanning is used when the holes exceed 2 in. dia. The material is removed as a solid core, not in the form of chips, as in gun-drilling. Stricter limits are achieved in diameter and straightness, holes are drilled to a greater depth, metal is machined away more rapidly, and as "discard" produces a core rather than a mass of chips, the core being of greater value.

Tools are rigidly held and carefully set-up to prevent misalignment, inaccuracy to dimensions and finish, and shorter service-life of the tool.

Machines for deep-hole trepanning may be engine-lathes, turret-lathes or horizontal drillers. The work is always rotated, the tool being fixed. The tools may be hollow boring-bars with solid carbide or carbide-tipped cutters. Single-cutter trepanning heads are self-piloting and guided by wear-pads placed about 90–180 deg. behind the cutter. The wear-pads balance the cutting-power, regulate the diameter and dimensions of the hole, and have a burnishing effect on the surface. They usually have bodies of steel on to which carbide wearing-surfaces are brazed.

The start of the cut necessitates use of a guide bush, counterbore or grooving tool. Ordinary trepanning cutters have a carbide tip brazed onto a body of tool steel. Multiple-cutter heads are used if starting is troublesome or a degree of shock is probable.

Speeds of deep-hole trepanning range from 30–110 surface ft./min. for carbon and low alloy steels of ordinary composition, and 30–135 for free-machining carbon and low alloy steels, using high speed steel cutters. For carbide-tipped cutters, the corresponding speed ranges for both materials are 350–600. Sulphurized cutting-oils are best for deep-hole trepanning as long as the supply is constant and controlled enough to remove the chips.

Beryllium, heat-resisting steels and most other classes of steels can be trepanned.

Tripod-punches In locating holes to be drilled, it is usual to employ a tripod-punch instead of the standard centre-punch because there is then no work-hardening of the spot marked, which might have a harmful effect on the drill-point when it makes contact. This is particularly necessary when stainless steels are drilled.

True Rake See **Tool Angles.** An angle measured at 90 deg. to the projection of the cutting-edge on the reference-plane. It is the resultant of the radial, axial and corner angles, and gives a fair idea of the eventual efficiency of a cutting-tool. The true rake influences the tool service-life, the type of surface-finish of the work, the amount of power required and the stresses on the machine tool.

Truing Grinding-wheels Changing the form of an abrasive wheel to generate a specific contour or to maintain the desired accuracy of contour. **Dressing** is a modification of the wheel's cutting action by a sharpening process, but the word truing is often used as if it meant dressing, and vice versa. A hard steel corrugated disc; abrasive material in the form of a revolving wheel; or mounted diamonds, presented at an angle of 60 deg. to ensure efficient cutting with minimum heat generation and maximum tool service life, are used. Truing may be done with **crush-trued** wheels (q.v.) using a grinding-oil to true multi-ribbed abrasive-wheels. Diamond-truing of these wheels is used on resinoid bonded types, and to produce fine, accurate pitches.

As long as the wheel is not glazed by excessive work, it needs only a fine finish, and one truing should keep it in good cutting condition for a long time unless the work is particularly exacting. The diamond-cuts are from 0·0008–0·0012 in. deep for finishing, with a fine traverse and plentiful water as a coolant. For roughing, the maximum feed is 0·001 in. using a single pass. Heavier cuts are made with a special wheel-dresser or other tool, the diamond serving as a finishing-tool. Cuts that are too deep may force the diamond out of its setting.

The wheel is trued by being run at the working speed except in the cases of thread-grinding and specially large wheels, when a small reduction of speed may be allowed. The tool is usually mounted in a special hole in the tailstock. Feed is swift to produce a coarse surface-pattern, but slower for finishing. The diamond is set slightly below the centre-line, the centre-line of the tool shank making an angle of 10–15 deg. to produce a trailing-cut. In a more modern method the

TABLE XIV

Grinding-wheel Truing Times

	Roughing				*Finishing*			
Wheel width (in.)	2	4	6	8	2	4	6	8
Time (secs.)	30	45	60	75	60	90	120	150

diamond makes contact with the wheel at a point connecting the wheel's centre with that of a circle corresponding to the work. This is in effect the ultimate point of contact between wheel and work. The diamond is mounted in a fixed tool post and re-set when it has worn level with the metallic setting. Typical truing times are given in Table XIV.

T-slot Cutters A type of **milling cutter** (q.v.) used for forming the tee-shaped slots in work-tables, the tool being used in enlarging the bottom of the slot after it has been milled in a milling machine. The cutters are usually of standard dimensions and either right- or left-hand, of a size conforming to variously-sized bolts. They range in diameter from $\frac{1}{4}$ to $1\frac{1}{2}$ in. and are made in standard sizes.

T-slot mounting A principle of mounting used mostly when tools work from a cross slide, irrespective of the type of tool adaptor used. The mounting is additionally for fixing adaptors to carriers travelling axially on main-slides of multiple-spindle machines. The tees resemble those for cross-slide use, but are parallel with the tool-cutting direction, which is axial, *not* at an angle of 90 deg. to the cutting direction.

Tungsten-carbide Tools See **Carbide Tools.**

Tungsten-copper Electrodes Sintered electrodes for **electro-chemical machining** (q.v.), composed of 80 per cent tungsten with 20 per cent copper. Machined copper-tungsten electrodes are also used in electrical discharge machining (q.v.), but not as often as are graphite electrodes.

Turning Machining the external surface of a component or a rotating face in a **lathe** (q.v.) or the internal surface by **boring** (q.v.). The work is rotated at a speed high enough to give the greatest economy in machining, and a tool is presented to it which, by the exertion of power, causes the cutting-edge to bite into the metal to a specified depth termed the "cut". At the same time the tool is moved sideways along the length of the work, this movement being termed the "traverse". As both movements are simultaneous, there is a slight sideways tool-traversing motion each time the work turns one

Copy turning with increasing diameters   Single-point turning and facing   Boring and facing

Fig. 59   Turning operations

revolution. The distance the tool travels for each revolution is termed the "feed".

Turning-tools Lathes for turning, whether capstan- or turret-, use a rotating toolholder carrying a range of tools all adjusted for their work, and brought into action one after the other so that different operations can be carried out on an individual component or bar, whereas in ordinary lathes only one tool can be used at a time. The modern capstan-lathe or turret-lathe (q.v.) can machine with up to 10 or 12 tools, each performing a different function, cutting to fine limits at high speeds, so economizing considerably in production-time.

Turning-tools rough off the metal at high speed. The straight-nosed roughing-tool (see Fig. 53, page 182) is characteristic for this operation, but a curved-nose roughing-tool is also suitable, and is said to give easier cutting with less power-consumption. Tools with round or curved noses enable larger rake angles to be used without causing chatter, and produce small, rather than long, spiral chips. They are often bent or cranked to enable them to machine in the neighbourhood of a shoulder that would obstruct the toolpost or holder of a straight tool.

Allowable speed of tool becomes greater as the nose-radius increases, particularly if in making shallow cuts the tool most influential on speed, and the speed itself, are in inverse relation to the feed. Speeds for tools tipped with tungsten carbide are about three times those for high speed steels for the same operation where carbide tools can be used. Finishing-speeds are comparatively high even when 0 deg. side rake and side-cutting edge-angles give a smooth surface.

Turret-drilling Machines Just as there are **turret-lathes** (q.v.), so there are drilling machines with a turret of cylindrical type carrying separate spindles. The machines are of two kinds: (a) that in which the turret rotates horizontally; (b) that in which it rotates vertically. The direction in which the drills are presented to the work corresponds to the direction of rotation. These machines are used where more than one drilling operation is required. The first tool in the turret may bore a hole to a prescribed diameter with a twist drill. The turret is then rotated to present the following tool, which may be a reamer or counterbore for bringing the hole closer to highly accurate dimensions; alternatively holes of different diameters are drilled at different levels in work held vertically, e.g. a hollow shaft having holes at 90 deg. to the axis.

The machine-spindle is bored-out to secure the drill-tang, or a socket may be attached to the spindle and the drill locked into this. This is the more usual method. Direct insertion of the drill in the spindle is largely restricted to straight- or parallel-shank drills. Chucks may also be used to secure the drill, and this method is becoming more popular, especially for small drills or those with straight shanks. These chucks can accept a range of drills of varying diameters, whereas otherwise a separate socket has to be used for each drill diameter. Taper-shank drills are not usually held in chucks or collets, but chucks have been designed to take both taper-shank and straight-shank. Adaptors are also used to enable a chuck for straight-shank to take a taper-shank drill.

When taper-shank drills are used in sockets, the drill-end is made somewhat flat, and locks in a corresponding groove or slot inside the socket so that the drill cannot slip round when working. See also **Drilling.**

Turret-lathes In turret-lathes the revolving toolholder is directly supported by the lathe-saddle moving along the lathe-bed, the length of which bed determines the length of workpiece that can be machined. Complicated machining operations are achieved by rapid and accurate repetition (see **Lathes**). The slide is taken back at the completion of the cut, the entire saddle and apron being moved by a quick power-traverse in either direction. This form of lathe gives the maximum rigidity for heavy cutting.

When the work involves a greater number of tools than the turret can take, a cross-slide is added. Usually the turret takes eight tools, an additional five being carried by toolposts on the cross-slide. Still more tools can be accommodated in knee-type toolholders. All these tools or any individual one cut simultaneously, so that the machine produces work much faster than can ordinary lathes. Two common means of indicating the size of a turret-lathe are used. For working on bar material fed through the machine-spindle, the size given by the makers shows roughly the greatest possible diameter that will pass through the spindle and length that can be turned. Alternatively, in a turret-lathe for chucking, the given size represents the greatest possible diameter the machine will swing over the lathe-bed ways.

Twist Drill A cutting-tool having two helical flutes opposed to each other and covering the entire length of the tool. These flutes give the correct cutting angle on each portion of the drill length irrespective of any regrinding. The drill-point is formed by two surfaces similarly shaped and at an angle to the drill-axis. The form and position of these determine the clearance angle of the drill. (See **Tool Angles.**) The parallel planes of the cutting lips or edges are separated by a thickness of metal termed the **web.** Regrinding is always confined to the point.

The chisel-edge makes the first cut, which is rapidly enlarged and deepened by the flutes. The non-cutting end of the drill is either cylindrical, with a flat base, or tapered with an end flattened into a tang to engage with a socket. The first type are termed straight-shank, the second taper-shank, drills. The straight-shank drills are smaller in diameter and range up to about $\frac{1}{2}$ in., whereas taper-shank drills are mostly above $\frac{1}{2}$ in. Modified types can be obtained (see **Jobbers' Drills, Wire Drills**, etc.). Both English and metric sizes are made at present.

Efficiency of these tools is largely established by the method of securing the drill in the machine (see **Turret-drilling Machines**). They are run at the highest speed that will not dull the cutting lips, and reground after an economic amount of work is done. The drills are made of tungsten carbide for porcelain, glass, and some plastic materials; but most drills for metals are made of tungsten high speed steels or, for the most difficult and stubborn materials, high speed steels with a relatively large cobalt content. Butt-welded drills with the fluted portion electrically welded end-to-end to the plain shank have in England largely replaced solid drills.

Drill-performance has been improved by shortening the cutting-flute length to give greater rigidity and prevent fractures and badly-drilled holes. The chisel-edge is also shortened to lessen the tool-thrust or cutting-pressure, so improving its cutting-power. Well-designed grinding-machines ensure that both cutting-flutes cut equally and the original tool angles are maintained. Drilling-speeds range from 80 to 100 ft./min. for mild steel and malleable iron; 50–70 for alloy and tool steels; 200–300 for non-ferrous metals; 100–150 for cast iron and plastics. The feed is about 0·001 in. for each $\frac{1}{16}$ in. of drill diameter/revolution, for high speed steel drills only. 50 per cent less is needed for carbon steel drills.

For all metals except cast iron and high manganese steel (11–14 per cent manganese), which are drilled dry, a soluble cutting-oil is used as a cutting-fluid.

Two-die Thread-rolling Machines In **Thread-rolling** (q.v.), machines embodying two dies of larger diameter than those in three-die machines, the diameter being pre-determined by the machine-size and that of its fixtures. Thus, a machine taking parts $\frac{1}{16}$–$1\frac{1}{2}$ in. employs dies 5–6 in. dia. for all work-dimensions. The minimum work-diameter for these machines is 0·05 in., and the maximum depends on what equipment is at hand, but is at least 6 in. dia. by 12 in. long.

Two-lip Gun-drill A type of drill for drilling the forged steel barrels of rifles. It has two flutes running longitudinally to the drill-axis, polished on their faces to ensure ready clearance of chips. The chisel-point is precisely on the centreline of the drill, and the lips are notched on their ends with staggered grooves for breaking up the chips. The drills are claimed to give three times the output of a standard drill on this class of work. The back of the drill is taper-ground, and the tip is round.

Two-plate Machine Lapping Mechanical **lapping** in which the laps are two cast iron plates or abrasive circular plates using a carefully bonded abrasive. They are normally 8–28 in. dia., but larger diameters are obtainable, and have mostly plain faces whose width is no greater than that of the work surface, to ensure maximum precision. The laps are secured by vertical machine-spindles. The work is held between the plates in slotted discs and given a rotating and sliding motion. It is also given an eccentric or in-and-out motion to keep inner and outer edges in contact with the lap and eliminate grooves on it. This is not needed if the output required is small and the laps have been reconditioned to keep them fully flat.

Both sides of the work are lapped simultaneously when abrasive lapping is done, and the lap-flatness is ensured by diamond-dressing. Carrier-rings revolve as the laps turn, giving the work a cycloidal movement as it passes over the laps. Loose abrasive is employed.

U

Ultrasonic Machining A comparatively modern process in which ultrasonic sound waves, developing vibrational energy, are used to take off material with the aid of abrasive particles vibrating in a slurry circulating through a narrow gap between the work and a tool oscillating at about 20,000 cycles/sec. The tool develops its own shape in the work with an accuracy governed by the tool-size, machine-robustness, tool-rigidity, slurry-temperature, size of abrasive particles and method of operation. The process is primarily applied to material low in ductility, but of considerable hardness, and either metallic or non-metallic materials can be machined.

The vibrational energy developed is transferred to the water slurry and concentrated in the area through which the work passes.

UNDERCUTTING

Essential to the operation are an efficient electrical generator and a **transducer**, or means of interchanging electrical energy and mechanical motion.

The operation, extremely expensive, is employed only when the intricacy of the work contour makes other machining operations impracticable. The parts to be machined must not be electrically conductive. Although many inductile parts are machined in this way, ductile metals also can be treated. Honeycomb materials in particular lend themselves to this type of machining. The rate at which metal is machined off is low, but comparatively shallow uneven cavities can be produced.

The machine consists usually of a coil-lead from an oscillator and amplifier, connected to a magnetostrictive stack, united by a connecting body to a tool-holder carrying the tool. The work is immersed in the abrasive slurry and held by a suitable fixture. A pump keeps the slurry in circulation and a refrigerating system cools the slurry to between 1·7 and 4·4 deg. C. (35–40 deg. F.).

The abrasives are silicon carbide, boron carbide and aluminium oxide, boron carbide being the most popular. The slurry mixture is usually 30–60 abrasive by volume + water, though larger tools require a lower abrasive content than smaller. Grit-size is 200–400 for roughing, 800–1,000 for finishing. The tool is usually of low carbon or stainless steel and formed to the type of contour required, being brazed or soldered to the tool-holder. Steel is not economically machined if its hardness is less than Rockwell C45. Typical applications of the process are to tool steels, stainless steels, heat-resisting steels, germanium, glass, ceramics, carbides and semi-conductors.

Containers for the slurry are needed.

Undercutting Relieving the teeth of tools such as those used in broaching, carried out in either the **lathe,** the **grinding machine** or by **chemical blanking.** It may be combined with other operations, especially in lathe work. (See **Boring Machines.**)

Underpass Shaving See **Shaving.**

Universal Chuck A concentric type of chuck much employed in **engine lathes** (q.v.) owing to the simultaneous motion of the chuck jaws. Work of circular form is rapidly seized and held in a position concentric to the lathe-spindle. The chuck jaws maintain themselves equidistant from the lathe-centre and are adjustable by a screw or screws, being moved by gears working in a common rack or scroll or both combined. Alternatively, the movement can be made manually or by key. The chuck itself comprises front and back plates or an inner or external shell in which the gearing is enclosed. The universal chuck enables circular or other work to be held centrally for turning without adjustment, but the work must be uniform in shape. It is sometimes termed a **self-centring chuck**.

Universal Head See **Boring Tools**.

Universal Knee-and-column Milling Machine A milling machine in which a two-piece saddle allows the machine-table to be pivoted for cutting radially. It embodies a dividing-head and a special mechanism for driving. The majority of these machines also include a detachable head fixed to the pillar or column for use when the machine operates a vertical spindle. In some types the head is pivoted through an angle of 180 deg. The machine can be operated automatically. See **Milling Machines**.

Universal Thread-milling Machines Machines for cutting screw-threads (see **Threading**) in which are embodied a lead-screw and change-gears enabling leads of $\frac{1}{32}$–60 in. to be used. The machines cut all types of internal and external threads, but not square threads. An extensive speed-range is obtainable and the cutter head can be set to give the angle for either right- or left-hand of thread-helix. The cutter is of single form and traverses the thread length at this angle.

Up Milling See **Climb Milling**.

V

Vacuum Chip Remover In machining beryllium, a vacuum pipe situated close to the operating region of the cutting tool for the removal of chips in turning cylindrical billets of the commercially pure metal.

Vacuum Clamping A clamping system in which the workpiece is firmly held down with a vacuum produced by an electric pump.

Vanadium Carbides Free grains of vanadium carbide embodied in high speed steels, introduced into the composition to give the steels a greater abrasion resistance and on occasion a longer tool service-life. The drawback is that they make the steels somewhat more difficult to grind.

Vehicle for Lapping A fluid in which abrasive particles are suspended for lapping. It may be grease, oil, water or a volatile spirit such as alcohol, depending on the purpose. The vehicle of the heaviest body usually gives the best finish, with a smaller amount of material removed for each pass. When maximum cut is the essential need, alcohol is used, but gives a less attractive finish.

Vehicles for lapping must hold the abrasive in suspension irrespective of conditions, and must be little if at all affected by changes of temperature. They give the greatest possible cut with the highest quality of finish; reduce frictional heat; have no corrosive effect on the work surface; are non-toxic and non-injurious to the human skin; and need no dangerous cleansing agent for their removal. (See **Lapping.**)

Vernier A device for measuring in minute fractions a distance previously known approximately. This device was invented by the Frenchman Pierre Vernier over 300 years ago. It consists in essence of two parallel facing scales; a short sliding scale divided by equally spaced marks greater or less by 1 in number than the graduations covering the same length on the main scale. The vernier is applied in principle to various gauges and calipers, protractors, and many slides of machine tools. Mostly the scale readings are to 1/1000th in.

Vertical-boring Mills Machine tools fulfilling the basic purposes of the lathe, but better suited than the lathe to large, heavy work such as big rings and short cylinders. The work is positioned on a horizontal table, set up, made level, and clamped down. Counterbalance may be given if the load is off centre. Two tools or more are used in combination to perform boring and turning either together or in succession. The machine needs less floor area than does a lathe of corresponding capacity. (See **Boring.**)

Vertical-broaching Machine A machine in which the broaching tool is held in a driving-carrier running in a single slide, the cutting-fluid being taken through a pipe to the work. Start and stop controls lie within easy reach of the operator. The work is held in a quick-acting clamp or jig. The massive vertical body holds the hydraulic mechanism and driving motors. The clamp or jig is fixed on a horizontal work-table automatically moved in to the correct position for machining and moved out again after completion of cutting. As soon as the work-table is in position for broaching, a valve automatically opens and the broach-slide moves down, carrying the broach with it. This slide then reverts to its original position. (See **Broaching.**)

Vertical Drill-press See **Drill-press.**
Vertical-lapping Machine See **Lapping Machines.**
Vertical-milling Machine See **Milling Machines.**
Vertical-spindle Surface-grinder A machine embodying a powerful built-in drive to the wheel-spindle, continuously-variable

VIBRATION DAMPERS

hydraulic table-speeds, with run out and stop loading-positions, no reversal shock irrespective of speed, permanent protection and automatic lubrication of table ways, sensitive vertical adjustment and quick hand-motion of the grinding-wheel head, push-button motor-control, a means of separating out the swarf, efficient guards for water and wheel, and a high capacity for removing metal.

The machine may have either reciprocating or rotary work-tables; for the most part the reciprocating tables employ cup, cylinder or segmental grinding-wheels. The column supporting the wheel-head may be fixed or laterally sliding on ways to cope with work of greater width than the wheel. The rotary-table machines employ cylinder or segmental wheels, sometimes up to 5 spindles being mounted on a central column. A ring-type rotating table then carries the work under the wheels. Roughing-, semi-finishing- and finishing-cuts are taken on such parts as engine blocks with a single pass.

These vertical grinding-machines are particularly serviceable when two opposing flat surfaces need to be ground. Large or small outputs can be handled, and tolerances of 0·001 in. and surface-finish not more than 32 micro-in. are attainable. Size limits are ±0·0002 in. and parallelism 0·0001 in. or more. Surface finish attainable is less than 10 micro-in.

Vibration Dampers See **Damper Plugs.**

Vitrified-bond Grinding-wheels Wheels used in grinding-machines for grinding the surfaces of metal parts, made up of abrasive particles held together by a bond consisting of a fused ceramic material manufactured in a kiln at a high temperature and not subject to change by temperature. The wheels are made considerably stiffer than by other bonds and are therefore less likely to distort, so that maximum accuracy is achieved. Accurate contour is also obtained when the wheel is trued. Progressive error in the long lead in the threads cut in thread-grinding is also unlikely, and therefore little trouble is found in rectifying previously-cut threads with some error in the lead. The vitrified bond wheels are therefore used for fine limits of contour or lead, lead-screws, threading-taps having to be finish-ground to great accuracy, gauges, worm gears in which a small feed is combined with numerous cuts, internal threads, etc. On the other hand the wheels are somewhat brittle, and must therefore be carefully handled. The bond is without elasticity, but can be produced with greater hardness-range than other bonds. Its only limitation is that the wheel-thickness may not always withstand a lateral pressure beyond its capacity. (See **Grinding.**)

W

Wavy-set Tooth Pattern In setting the teeth of contour band saws (q.v.) when thin cross-sections have to be cut, the teeth are gradually offset first to the right, then to the left (see Fig. 50, page 173), the pattern so formed resembling a wave. This wave enters the metal more gradually and uniformly, so that there is the minimum shock-loading of the teeth.
Wax Cutting-fluids Water-soluble waxes sometimes used in the machining of titanium alloys. (See **Cutting Fluids**.)
Web The central part of a twist drill body uniting the **lands** (q.v.) and **flutes** (q.v.). The front portion is ground to form the cutting-point or -edge and slowly thickens in the direction of the shank. Its thickness is normally measured at the drill-point, and it gives support to the drill as it penetrates.
Wheels See **Grinding, Polishing,** etc.
Wheel Swarf See **Swarf**.
Whitworth Thread Standard forms of angular section screw-threads of both British and American types.
Wickman Gauge An adjustable gauge of horseshoe form for rapid use.
Wire Drills Drills of diameter so small that they have to be produced from lengths of suitable steel wire, usually of high-quality carbon steel, to standard wire gauges.
Wire Gauge A notched plate provided with gauged slots, each carrying a number corresponding to the sizes of wire produced. The respective gauges of this type are often referred to by initials, e.g. S.W.G. (Imperial Standard Wire Gauge), B.W.G. (Birmingham Wire Gauge).
Woodruff Cutters Standard cutters for the production of standard Woodruff keyways. (See **Milling Cutters**.) Each key-dimension requires an individual cutter given a number of letters corresponding to that key. They are made both right- and left-hand with parallel-shanks which taper slightly as they approach the cutting-edges to ensure a cut of greater depth. Their diameters range from $\frac{1}{2}$ to $1\frac{1}{2}$ in., the sides being ground to give a degree of clearance, and the cutting-face on the circumference only. Their teeth may be fine or coarse.
Wool-felt Polishing-wheels See **Polishing** and **Felt Polishing-wheels**.

Work Horn In broaching (q.v.) a guide to locate and guide the broach when cutting keyways. It is, in fact, a special kind of face-plate. The ground diameter on the back of the horn beds closely into the platen of the machine or into a reducing-bush, according to the dimensions of the work. A round pilot on the opposite end, 0·001 in. smaller than the hole in the work, gives exact work location. The slotted horn guides the broach, and the keyway depth broached is governed by the slot-depth and the height of the final tooth on the broach. The horn is hardened and ground to size on all working-surfaces. The work is held square and parallel to the axis of broach-travel. The pull of the broach through the cut keeps the work securely in place.

Wringing Together See **Gauge Blocks.**

Z

Zero Backlash Gear Honing In this technique commercial gears are honed by locking the tool-head to render constant during the entire cycle the distance between work-centre and honing-tool centre.

Appendix: Theory of Machining Outlined

In order to simplify the initial stages of this brief outline we shall first consider the simple wedge tool, with a straight cutting-edge that is parallel to the original plane surface of the workpiece and perpendicular to the direction of cutting, with a cutting-edge of greater length than the width of chip removed. If we have clearance α, wedge angle β, then rake χ is defined as follows:

$$\chi = 90° - (\alpha + \beta).$$

Hence, if $(\alpha+\beta)$ exceeds 90° the rake is negative, and if $(\alpha+\beta)$ is less than 90° the rake is positive. Rake affects chip formation, chip type, the degree of smoothness of the machined surface, and the rate of tool wear; clearance too affects tool wear.

When a simple wedge tool is in operation as described above, the process is termed *orthogonal cutting*.

Types and Formation of Chips

There are three basic types of chip, described as tear-, flow- and shear-type, though these terms should not be taken as accurate descriptions of the processes involved in their formation.

Tear-type: As the tool traverses the workpiece severe distortion of the metal in close proximity to the cutting edge occurs, with the eventual result that a crack is formed, preceding the tool and separating the distorted metal from the comparatively unaffected workpiece. Due to increasing stress, the chip eventually breaks away.

As in general the crack tends to run slightly inwards into the work surface, there is a greater depth of cut than that required.

Flow-type: With flow-type chips there is continuous cutting, and chip formation does not normally result in chip fracture.

Of interest is the formation of a *built-up edge*: this is a highly-deformed area of metal seated on the tool face, and it participates in the cutting process. The built-up edge will be discussed in greater detail later.

Roughly speaking, the severe plastic deformation of the workpiece causes the metal to shear along a series of planes, which build up, one on top of the other, to form the chip. It is important to realise, however, that the shear-planes are not normally separated from each other.

Shear-type: Cutting is in this instance, as with the tear-type, discontinuous. There is severe plastic deformation of the metal adjacent to the tool face in the same way as there is with the tear-type chip, but there is also a band of deformation running between the cutting edge and the original workpiece surface, and fracture may result along this band.

The Machined Surface

Because the metal close to the machined surface has been severely plastically deformed, this area is much harder than the rest of the workpiece, and the phenomenon is known as *work-hardening*. Regions of greatest hardness are, in decreasing order, the built-up edge, the part of the chip nearest to the cutting edge, the chip, and the material just ahead of the cutting edge adjacent to the machined surface.

The degree of the deformation, and hence the extent of work-hardening, increases with increasing negative rake, and is proportional to the chip thickness. With the exception of very low rates of progress, where the built-up edges may cause anomalous high deformation in a localized area, cutting speeds have apparently little effect on work-hardening.

There are also residual sub-surface stresses which, it is thought, have an effect upon the fatigue resistance of the part concerned. These may possibly be so severe as to cause cracking of the machined part.

Lubrication

It is thought that the formation of chips is largely governed by the degree of friction between chip and tool face. This is not the place for a detailed discussion of the nature of friction in this circumstance; it is sufficient to note that it is usually considered to be the sum of two effects: (1) the force required to drag the projections of one rough surface through the projections of the other; and (2) the force required to shear the intermetallic junctions in areas where the two surfaces are touching. The coefficient of friction increases with increasing rake angle, rather than remaining constant for any single combination of materials used for tool and workpiece (when cutting dry).

Cutting fluids, in many cases, assist the efficiency of the cutting operation by causing a chemical reaction with the newly-exposed (and hence uncontaminated) surface of the chip in contact with the tool face. An organic lubricant may react with the work surface to produce a compound that has low shearing strength, thus decreasing the friction caused by (2) above. For example, carbon tetrachloride (CCl_4) reacts with an aluminium workpiece to produce

APPENDIX

aluminium chloride ($AlCl_3$), a compound with extremely low shearing strength.

It should be noted that the lubricant effect of cutting fluids decreases as cutting speed increases, and considerably reduces work done per unit of material removed when cutting speed is low.

Temperature

Although it is difficult to measure or examine the steep temperature gradients generated within the chip, the problem is not insurmountable, the technique used being to photograph the side of the chip using a plate sensitive to the infra-red radiations emanating from the chip surface, as the differing intensities of these indicate differing temperatures.

One can assume that all the energy used for cutting is eventually converted into heat energy, since the work done against friction is released as heat along the chip-tool boundary, and the work done to cause the sub-surface plastic deformation is released at the shear plane between the junction of chip with work surface and the junction of the machined surface with the cutting-edge of the tool.

The temperatures of these areas can be measured as described above, and the knowledge of their values is of considerable assistance in the studies of tool wear, type of chip formation, etc.

Chatter

The vibration of either cutting tool or workpiece (or both) is known as chatter, and may cause imperfections on the machined surface, high tool wear, fatigue failure of machined parts, and sustained rebarbative noise during the cutting operation. The vibrations are of two basic types:

Forced vibrations: During discontinuous cutting there is, of course, a periodic variation in the cutting force. The vibrations of the system have the same period as these cutting force vibrations, and, at frequencies close to the natural frequency of either tool or workpiece, resonance is set up, greatly amplifying the vibrations.

Self-excited vibrations: The most common example of a self-excited vibration is that of a violin or 'cello string. For roughly the same reasons, machine tools may tend to display vibration at frequencies around the natural frequency of the system. For a detailed discussion of self-excited vibrations, the reader is recommended to study any suitable text on the mechanics of oscillation.

Chatter is at its greatest in systems that incorporate broad thin chips, large cutting forces, and tools with a negative rake. The reduction of chatter can be assisted by the introduction into the system of a *damping agent*.

Roughness

It is naturally of interest to determine the optimum conditions for the production of a smooth surface by the cutting operation itself. There are various techniques of improving the resultant surface, such as: the consideration of the rake angle for various cutting speeds; lubrication; different tool materials; and decreased surface roughness of the tool faces. These are dealt with briefly below:

At high speeds roughness decreases with increasing negative rake, while at low speeds the roughness decreases with increasing positive rake. This is probably due to the effects of speed and rake together on the built-up edge, since a reduction in the size of this edge seems to reduce the size of the fragments of metal that adhere to the machined surface.

This decrease in roughness due to reduction in the size of the built-up edge renders lubrication another effective improvement. As discussed earlier, the friction of a lubricated system increases with increasing cutting speed, and so the effect of the lubricant in reducing the size of the built-up edge decreases at the same time: fortunately, however, increased cutting speed tends to decrease the size and hence the necessity for a lubricant.

For most systems there is a range of cutting speeds which are too rapid for a lubricant to be effective but not rapid enough to provide a sufficiently smooth surface. In such instances it is necessary to either increase the cutting speed, or to reduce it and make use of a lubricant, the technique adopted depending on any other factors that may have to be taken into consideration.

A change in tool material will affect the friction between tool and chip, and may hence alter the degree of roughness of the machined surface. In general, the lower the friction, the smoother the surface, and, so, for each material to be machined, it is advisable to consider carefully beforehand the material that the tool should be made of.

Surface roughness of the tool face has probably little effect on the degree of friction between chip and tool. However, roughness of the machined surface can be caused by a poorly finished tool, since a rough cutting edge tends to encourage the formation of a built-up edge.

The built-up edge

As the tool moves across the workpiece it elongates the material on the underside of the chip and on the surface of the workpiece. Under certain conditions, this elongated material will accumulate on the tool face to produce a built-up edge. With continued cutting the material at the junction of chip and workpiece is laid over the built-up edge and thus strained and work-hardened until fracture occurs.

In its turn, the built-up edge also fractures and, because of the extremes of temperature and pressure at the cutting edge, part of it may adhere to the workpiece, leaving the surface rough after machining. However, it appears that the base of the built-up edge remains permanently seated on the tool face and that only the upper part of it fractures: the base will then frequently become welded to the tool and possibly cause permanent damage to it.

The built-up edge may either increase or decrease tool wear, depending on the circumstances. In the latter instance, the decrease is probably due to the protection to the tool caused by the position of the built-up edge between chip and tool.

Tool wear

Tools may gradually decrease in size and weight or they may suffer from crumbling of the cutting edge due to fracture. (This latter phenomenon is known as *spalling*.)

It appears that the factors governing tool wear are not so simple as might be supposed, and that rules-of-thumb such as "the harder the tool, the greater the wear resistance" are frequently inapplicable. Wear is probably caused by chemical reactions between the tool and the workpiece under special conditions of temperature and pressure, and such a reaction has in fact been observed between a steel workpiece and a tungsten carbide tool material.

However, there are too many unknown factors for anyone to do more than postulate the causes of gradual wear,[1] though its effects and nature under varying circumstances can be experimentally examined, and hence efforts made to prevent it. Gradual wear occurs both on the tool flank and on the tool face, where *cratering* may take place. Cratering takes the form, as the name might imply, of the appearance of a crater just behind the cutting edge on the tool face. This crater will enlarge until it has undermined the cutting edge sufficiently for fracture to occur. It is not yet understood why, under certain circumstances, cratering should be more important than flank wear, while, under other circumstances, their relative importances are reversed.

[1] For a fuller discussion of the facts discovered and the theories advanced concerning tool wear the reader is advised to consult *Metal Wear: A Brief Outline*, by Eric N. Simons.

It has been shown by experiment that, for a constant size of cut, there is a proportionality between cutting speed and tool life (i.e., the cutting time until the tool fails). This can be represented as

$$T \propto V^{\frac{1}{n}} \text{ (positive root)} \tag{1}$$

where T = tool life
V = cutting speed

and n is a constant determined, as is the constant of proportionality, by experimental means and depending on the work and tool materials.

If the constant of proportionality is C, then (1) above may be restated as

$$VT^n = C \tag{2}$$

If the curve of this identity is plotted for the particular case with cutting speed V shown against tool life T then the *permissible cutting speed* may be found for a specified tool life. In other words, if it is required that the tool last for at least x hours before failure, a suitable maximum value for the cutting speed may be determined by examination of the plotted curve (or, of course, by substitution into (2)).

The depth of cut and the rate of feed also have an effect on the permissible cutting speed, and their effects are related so that, in general, a high feed coupled with a low depth of cut or a low feed with a high depth of cut will minimize tool wear.

The effect of rake on tool wear is rather more complicated than might at first be supposed. As rake increases then so does permissible cutting speed until at some point a maximum value is reached and permissible cutting speed begins to decrease with increasing rake. Quite why this is so is not yet understood, and the optimum rake for a particular operation must be found by the results of experiment rather than by theoretical means.

Tool life can be prolonged by the use of a cutting fluid, which cools the system and lubricates the chip, thus reducing the severity of the conditions under which the tool is working. If the cutting fluid is applied between the tool flank and the workpiece by such means as a high speed jet then tool life is once again prolonged.

Although the most accurate determination of the permissible cutting speed is obtained by actual experimentation on the shop floor, literature has been published that contains empirical formulae from which an approximate value may be obtained. However, though such literature may have its uses, it is difficult to determine the degree of accuracy with which the empirical results may be applied to existing conditions, and it is thus recommended that not too much reliance be placed upon them.

Conclusion

It would be impossible in the few preceding pages to do more than give the briefest of outlines of some of the more elementary principles of machining, but it is hoped that the reader will have been helped thereby in his understanding of the main body of the dictionary. Of the numerous books available on the subject, should the reader wish to study the theoretical side of the field further, three have been selected, though it must be borne in mind that omission from this list is by no means a reflection on the quality or otherwise of any book.

Modern Workshop Technology Part Two, edited by H. Wright Baker, third edition 1969, Macmillan & Co.
Metal Cutting Mechanics, N. N. Zorev, 1966, Pergamon Press.
Metal Wear: A Brief Outline, Eric N. Simons, 1972, Muller.